System-Theoretic Analysis and Optimization of a Novel Secondary Radar Concept for Precise Distance and Velocity Measurement

Doctoral Thesis
(Dissertation)

to be awarded the degree of

Doktor-Ingenieur
(Dr.-Ing.)
the German equivalent of a Ph.D. in engineering

submitted by

Dipl.-Ing. Sven Röhr
from Merseburg

approved by the
Faculty of Mathematics/Computer Sciences
and Mechanical Engineering,
Clausthal University of Technology,

Date of oral examination:	05.10.2009
Chairperson of the Board of Examiners:	Prof. Dr.-Ing. Norbert Müller
Chief Reviewer:	Prof. Dr.-Ing. Martin Vossiek
Reviewer:	Prof. Dr.-Ing. Dr.-Ing. habil. Robert Weigel

Bibliografische Information der Deutschen Nationalbibliothek

Die Deutsche Nationalbibliothek verzeichnet diese Publikation in der
Deutschen Nationalbibliografie; detaillierte bibliografische Daten sind
im Internet über http://dnb.d-nb.de abrufbar.

D 104, Dissertation Clausthal, 2009

ISBN 978-3-8325-2459-3

Logos Verlag Berlin GmbH
Comeniushof, Gubener Str. 47,
10243 Berlin
Tel.: +49 (0)30 42 85 10 90
Fax: +49 (0)30 42 85 10 92
INTERNET: http://www.logos-verlag.de

Abstract

In the last decade, the demand for precise location and tracking of people, vehicles, or objects in real-time has increased tremendously. There are many exciting applications with a huge market potential for wireless location solutions. Today, wireless location systems are widely used in both consumer and industrial applications.

This thesis focuses on the analysis and optimization of a novel secondary radar concept for precise distance and velocity measurement. The proposed system is comprised of two active radar stations, a base station and a transponder. It is shown how frequency-modulated continuous wave radar signals are utilized to synchronize the clock of the transponder to the clock of the base station with high precision. After synchronization the distance between the radar stations is measured similarly to the well-known frequency-modulated continuous wave radar principle. A novel extension of the measurement algorithm exploits the Doppler frequency shift of the radar signals to measure the relative velocity of both units as well. Furthermore, novel multiplexing schemes for multiple transponders are derived.

To verify the system concept at hand a measurement system was designed and set up within this work. A thorough mathematical analysis of the underlying algorithms for synchronization and distance and velocity measurement provides the basis for the identification of sources of error. The identified errors are then analyzed in detail. The most important sources of error include multipath propagation, the mismatch of the sweep rates of the radar stations, and the signal-to-noise ratio of signals involved in the measurement process. Theoretical calculations as well as simulation and measurement results are used to evaluate the effect of each parameter on the performance of the radar system at hand.

Finally, the performance of the measurement system is evaluated in various environments. The results of an extensive measurement campaign prove the excellent performance of the novel system concept investigated in this thesis.

Preface

The thesis at hand is the result of my research at Siemens AG in Munich and Symeo GmbH in Munich / Neubiberg in cooperation with the Institute of Electrical Information Technology at Clausthal University of Technology. I started out to work on the project back in 2005 at the Corporate Technology department of Siemens. Later on, I joined Symeo GmbH, a spin-off of Siemens founded the same year. I would like to thank both companies for funding my research and giving me the opportunity to work on a most interesting scientific project. Furthermore, I appreciate the opportunity to attend renowned conferences all over the world.

Most importantly, I would like to thank Dr. Peter Gulden for the countless discussions we had throughout the years. His questions, ideas and advice have proven to be a most valuable source of inspiration. I am very grateful to Prof. Martin Vossiek, who supported me throughout my entire research and provided me with new points of view when necessary. Furthermore, I would like to thank Prof. Robert Weigel, who took the position of a reviewer on my doctorate committee, and Prof. Norbert Müller, who chaired the committee. I owe special thanks to Dirk Becker for his invaluable advice on hardware design.

It would have been hard to finish this thesis without the help of and encouragement by my friends and family. My warmest thanks go to my parents, who supported me throughout my life. I am also grateful to Thomas Hardtke, Dr. Thomas Deckert and Dr. Thomas E. Zander, who have read earlier versions of this thesis and provided me with countless suggestions on how to improve the presentation of the results of my research. Dr. Sebastian Siegel provided telephone and Internet support on urgent matters.

I am also grateful to the team of the Siemens Student Program (TOPAZ). They brought me into Siemens to exactly the right place at precisely the right time. Furthermore, special thanks goes to the fabulous musicians of Bon Jovi and Silbermond. Their music has brought me through one crisis or another.

Finally, I would like to thank all of my friends and colleagues at Symeo, who supported me and contributed to the successful completion of this thesis. It has been - and still is - fun working with you.

Lossa, March 2010

Sven Röhr

To my parents

Contents

List of Figures

List of Tables

Abbreviations

1D	One-Dimensional
2D	Two-Dimensional
3D	Three-Dimensional
A/D	Analog-to-Digital
ACI	Adjacent-Channel Interference
ADC	Analog-to-Digital Converter
AGV	Automated Guided Vehicle
AOA	Angle-Of-Arrival
BS	Base Station
DC	Direct Current
DDS	Direct Digital Synthesizer
DSP	Digital Signal Processor
EIRP	Equivalent Isotropic Radiated Power
FCC	Federal Communications Commission
FDMA	Frequency Division Multiple Access
FFT	Fast Fourier Transform
FMCW	Frequency-Modulated Continuous Wave
FPGA	Field Programmable Gate Array
FSK	Frequency Shift Keying
GLONASS	GLObal NAvigation Satellite System
GPS	Global Positioning System
iFDMA	Intermediate Frequency Division Multiple Access
IFFT	Inverse Fast Fourier Transform
ISM	Industrial, Scientific, and Medical
LO	Local Oscillator
LOS	Line-Of-Sight
LPR	Local Positioning Radar
meas.	Measurement

MUSIC   MUltiple SIgnal Classification

NLOS   Non-Line-Of-Sight

PDA   Personal Digital Assistant
PLL   Phase-Locked Loop
ppm   Parts Per Million, $1\text{ppm} = 10^{-6}$

RSS   Received-Signal Strength
RTOF   Round-trip Time-Of-Flight
rx   Receive

SNR   Signal-to-Noise Ratio
sync.   Synchronization

TCXO   Temperature Compensated Crystal Oscillator
TDMA   Time Division Multiple Access
TDOA   Time-Difference-Of-Arrival
TOA   Time-Of-Arrival
TS   Transponder Station, transponder
TS1   Transponder in measurement channel 1
TS2   Transponder in measurement channel 2
TS3   Transponder in measurement channel 3
TS4   Transponder in measurement channel 4
TS6   Transponder in measurement channel 6
tx   Transmit

UWB   Ultra-WideBand

VCO   Voltage Controlled Oscillator
VCXO   Voltage Controlled Crystal Oscillator

Symbols

a_{dl}	Attenuation of a delay line.
a_i	Attenuation of the component of the received signal corresponding to multipath i.
a_{path}	Path loss.
a_{sa}	Variable attenuation of a step attenuator.
A	Amplitude of a signal.
$A_{bs,tx}$	Constant amplitude of the FMCW signal transmitted by the base station.
$A_{ts,lo}$	Constant amplitude of the locally generated FMCW signal in the transponder.
$A_{ts,rx}$	Constant amplitude of the FMCW signal received by the transponder.
α	Ratio of the sweep rates of the transponder and the base station.
b_{bs}	Bandwidth of the low-pass filtered mixed signal in the base station.
b_{ts}	Bandwidth of the low-pass filtered mixed signal in the transponder.
B	Bandwidth of a sweep.
B_1	Bandwidth of the first synchronization sweep.
B_2	Bandwidth of the second synchronization sweep.
B_{bs}	Bandwidth of the sweeps of the base station.
B_f	Full bandwidth of pre-synchronization sweeps.
B_l	Lowest bandwidth of pre-synchronization sweeps.
B_m	Medium bandwidth of pre-synchronization sweeps.
B_{meas}	Bandwidth of the sweep for distance measurement.
B_{ts}	Bandwidth of the sweeps of the transponder.
c_0	Speed of light in the vacuum of free space.
$const$	Constant value.
c_{ph}	Phase velocity of the radar signals, usually smaller than c_0.
χ_s^2	Test statistic of the χ^2 goodness-of-fit test.
d	Distance between base station and transponder.

d_{crit}	Critical distance between the radar stations: if exceeded either f_{up} or f_{dn} is aliased.
d_k	Distance between base station and transponder in measurement channel k.
d_{laser}	Distance measurements obtained by a laser ranging system.
d_{lpr}	Distance measurements obtained by the LPR.
d_{lu}	Position of the linear unit.
d_{max}	Maximum measurement range of the system.
d_{max}^A	Maximum distance between the radar stations before either f_{up} or f_{dn} is aliased.
$d_{max}^{A_1}$	Maximum distance between the radar stations if f_{up} is aliased first.
$d_{max}^{A_2}$	Maximum distance between the radar stations if f_{dn} is aliased first.
$d_{max,k}$	Maximum range in measurement channel k.
d_{max}^+	Maximum measurement range of the system limited by the sampling frequency.
d_{meas}	Measured distance.
d_{meas}^A	Measured distance if f_{up} and f_{dn} are aliased.
d_{off}	Offset in distance if f_{up} and f_{dn} are aliased.
d_{sync}	Distance between base station and transponder during the synchronization upsweep.
d_{ts1}	Distance between base station and transponder TS1.
d_{ts2}	Distance between base station and transponder TS2.
d_{ts3}	Distance between base station and transponder TS3.
$DFTW$	Delta frequency tuning word of the DDS.
$\delta_{clk,bs}$	Absolute deviation of the clock frequency of the base station from its nominal value.
δ_{clk}	Relative deviation of the clock frequencies of the transponder and the base station.
δd	Distance measurement error due to false synchronization.
$\Delta d_{nlos-los,min}$	Minimum difference in length between the NLOS and the LOS path required to neglect the error due to interference.
Δd_{lpr}	Change in the measured distance between two consecutive LPR measurements.
Δd_{laser}	Change in the measured distance between two consecutive laser measurements.
ϵ_r	Relative permittivity.
f	Frequency of a signal.

f_1	Frequency of the low-pass filtered mixed signal in the transponder during the first synchronization sweep.
f_1^D	f_1 when the Doppler frequency shift is considered.
f_1^G	Frequency of the low-pass filtered mixed signal in the transponder during the first synchronization sweep (general solution).
$f_{1,center}^G$	Average of f_1^G during the synchronization upsweep.
$f_{1,start}^G$	f_1^G at the beginning of the synchronization upsweep.
$f_{1,stop}^G$	f_1^G at the end of the synchronization upsweep.
$f_{1,0}^M$	f_1 corresponding to the LOS path.
$f_{1,1}^M$	f_1 corresponding to the first (shortest) NLOS path.
$f_{1,i}^M$	f_1 corresponding to NLOS path i.
$f_{1,l}^M$	f_1 corresponding to NLOS path l.
$f_{1,max}$	Maximum possible value of f_1.
$f_{1,min}$	Minimum possible value of f_1.
f_2	Frequency of the low-pass filtered mixed signal in the transponder during the second synchronization sweep.
f_2^D	f_2 when the Doppler frequency shift is considered.
f_2^G	Frequency of the low-pass filtered mixed signal in the transponder during the second synchronization sweep (general solution).
$f_{2,center}^G$	Average of f_2^G during the synchronization downsweep.
$f_{2,start}^G$	f_2^G at the beginning of the synchronization downsweep.
$f_{2,stop}^G$	f_2^G at the end of the synchronization downsweep.
$f_{2,0}^M$	f_2 corresponding to the LOS path.
$f_{2,1}^M$	f_2 corresponding to the first (shortest) NLOS path.
$f_{2,i}^M$	f_2 corresponding to NLOS path i.
$f_{2,l}^M$	f_2 corresponding to NLOS path l.
f_{bin}	Spacing of the centers of the bins of the FFT.
f_{bs}	Instantaneous frequency of the low-pass filtered mixed signal in the base station.
f_{bs}^G	Instantaneous frequency of the low-pass filtered mixed signal in the base station (general solution).
$f_{bs,lo}$	Instantaneous frequency of the locally generated signal in the base station.
$f_{bs,rx}$	Instantaneous frequency of the signal received by the base station.
$f_{bs,tx}$	Instantaneous frequency of the signal transmitted by the base station.
f_c	Center frequency of the sweeps.
f_{clk}	Clock frequency of a radar station.

$f_{clk,bs}$ Clock frequency of the base station.

$f_{clk,ts}$ Clock frequency of the transponder.

f_d Frequency component of the low-pass filtered mixed signal in the base station proportional to the distance between the stations.

f_{dds} Output frequency of the DDS.

$f_{d_{max},1}$ Frequency component of the low-pass filtered mixed signal in the base station corresponding to the maximum allowable distance to the transponder in measurement channel 1.

$f_{d_{max},2}$ Frequency component of the low-pass filtered mixed signal in the base station corresponding to the maximum allowable distance to the transponder in measurement channel 2.

$f_{d_{max},k}$ Frequency component of the low-pass filtered mixed signal in the base station corresponding to the maximum allowable distance to the transponder in measurement channel k.

f_{dn} Frequency of the low-pass filtered mixed signal in the base station during the measurement downsweep.

f_{dn}^A Alias of f_{dn} measured if $f_{dn} > f_s/2$.

f_{dn}^{calc} Calculated frequency of the low-pass filtered mixed signal in the base station during the measurement downsweep.

f_{dn}^D f_{dn} if the Doppler frequency shift is considered.

f_{dn}^G Frequency of the low-pass filtered mixed signal in the base station during the measurement downsweep (general solution).

$f_{dn,center}^G$ Average of f_{dn}^G during the measurement downsweep.

$f_{dn,start}^G$ f_{dn}^G at the beginning of the measurement downsweep.

$f_{dn,stop}^G$ f_{dn}^G at the end of the measurement downsweep.

$f_{dn,k}$ Frequency component of the low-pass filtered mixed signal in the base station corresponding to the signal received from transponder k during the measurement downsweep.

$f_{dn,0}^M$ f_{dn} corresponding to the LOS path.

$f_{dn,1}^M$ f_{dn} corresponding to the first (shortest) NLOS path.

$f_{dn,i}^M$ f_{dn} corresponding to NLOS path i.

$f_{dn,l}^M$ f_{dn} corresponding to NLOS path l.

f_D Doppler frequency shift

$f_{fsk,0}$ Frequency of the transmitted signal if FSK symbol '0' is transmitted.

$f_{fsk,1}$ Frequency of the transmitted signal if FSK symbol '1' is transmitted.

$f_{fsk,lo}$	Frequency of the locally generated signal to receive and decode FSK data.
f_l	Lower frequency limit of the sweeps.
f_{rf}	Output frequency of the VCO.
f_s	Sampling frequency used to digitize the low-pass filtered mixed signal.
f_{ts}	Instantaneous frequency of the low-pass filtered mixed signal in the transponder.
$f_{ts,lo}$	Instantaneous frequency of the locally generated signal in the transponder.
$f_{ts,rx}$	Instantaneous frequency of the signal received by the transponder.
$f_{ts,tx}$	Instantaneous frequency of the signal transmitted by the transponder.
f_u	Upper frequency limit of the sweeps.
f_{up}	Frequency of the low-pass filtered mixed signal in the base station during the measurement upsweep.
f_{up}^A	Alias of f_{up} measured if $f_{up} < 0$.
f_{up}^{calc}	Calculated frequency of the low-pass filtered mixed signal in the base station during the measurement upsweep.
f_{up}^D	f_{up} if the Doppler frequency shift is considered.
f_{up}^G	Frequency of the low-pass filtered mixed signal in the base station during the measurement upsweep (general solution).
$f_{up,center}^G$	Average of f_{up}^G during the measurement upsweep.
$f_{up,start}^G$	f_{up}^G at the beginning of the measurement upsweep.
$f_{up,stop}^G$	f_{up}^G at the end of the measurement upsweep.
$f_{up,k}$	Frequency component of the low-pass filtered mixed signal in the base station corresponding to the signal received from transponder k during the measurement upsweep.
$f_{up,0}^M$	f_{up} corresponding to the LOS path.
$f_{up,1}^M$	f_{up} corresponding to the first (shortest) NLOS path.
$f_{up,i}^M$	f_{up} corresponding to NLOS path i.
$f_{up,l}^M$	f_{up} corresponding to NLOS path l.
F_f	Equivalent noise bandwidth of a window function.
FTW	Frequency tuning word of the DDS.
FTW_0	Frequency tuning word corresponding to the lower frequency limit of the sweeps.
FTW_1	Frequency tuning word corresponding to the upper frequency limit of the sweeps.
δf	Estimation error of Δf.

Δf	Offset in frequency of the sweeps of the base station and the transponder.
$|\Delta f|_{max}$	Maximum allowable offset in frequency.
Δf_a	Additional offset in frequency of the local synchronization sweeps in the transponder.
$\Delta f_{a,dn}$	Additional offset in frequency of the measurement downsweep transmitted by the transponder.
$\Delta f_{a,dn_1}$	Additional offset in frequency of the measurement downsweep transmitted by transponder 1.
$\Delta f_{a,dn_2}$	Additional offset in frequency of the measurement downsweep transmitted by transponder 2.
$\Delta f_{a,dn_k}$	Additional offset in frequency of the measurement downsweep transmitted by transponder k.
$\Delta f_{a,meas}$	Additional offset in frequency of the measurement sweep transmitted by the transponder.
$\Delta f_{a,up}$	Additional offset in frequency of the measurement upsweep transmitted by the transponder.
$\Delta f_{a,up_1}$	Additional offset in frequency of the measurement upsweep transmitted by transponder 1.
$\Delta f_{a,up_2}$	Additional offset in frequency of the measurement upsweep transmitted by transponder 2.
$\Delta f_{a,up_{k-1}}$	Additional offset in frequency of the measurement upsweep transmitted by transponder $(k-1)$.
$\Delta f_{a,up_k}$	Additional offset in frequency of the measurement upsweep transmitted by transponder k.
Δf^D	Estimate of Δf if both stations move relatively to each other.
Δf^E	Estimate of Δf including estimation errors.
Δf_p	Spacing of two peaks in the power spectral density.
$\Delta f_{p,min}$	Minimum spacing of two peaks to neglect the error due to interference.
Δf_r	Offset in frequency equivalent to the remaining offset in time after synchronization Δt_r.
Δf_{rf}	Quantization of the output frequency of the VCO.
φ_0	Arbitrary constant phase term.
φ_1	Arbitrary constant phase term.
φ_2	Arbitrary constant phase term.
φ_3	Phase term containing all terms that do not depend on t.
$\varphi_{dn,0}^M$	Phase of the component of the low-pass filtered mixed signal corresponding to the LOS path.
$\varphi_{dn,1}^M$	Phase of the component of the low-pass filtered mixed signal corresponding to the first (shortest) NLOS path.

φ_{ts}	Phase of the low-pass filtered mixed signal in the transponder.
$\varphi_{ts,lo}$	Phase of the local signal in the transponder.
$\varphi_{ts,rx}$	Phase of the signal received by the transponder.
γ	Signal-to-noise ratio.
γ_e	Effective SNR, determined by either the phase noise of the PLL or the receiver noise (depending on the attenuation of the signals).
i	Integer index.
I	Upper limit of i.
j	Imaginary unit, $j^2 = -1$.
k	Integer index.
K	Upper limit of k.
l	Integer index.
λ	Wavelength of the radar signals.
m	Integer index.
m_p	Position of the peak detected in the spectrum of the initial real valued FFT.
m_z	Position of the peak detected in the spectrum of the complex valued zoom FFT.
$m_{z,ip}$	Interpolated position of the peak detected in the spectrum of the complex valued zoom FFT.
μ	Sweep rate, ratio of the sweep bandwidth B and sweep duration T.
μ_1	Sweep rate of the first synchronization sweep.
μ_2	Sweep rate of the second synchronization sweep.
μ_{bs}	Sweep rate of the base station.
μ_{dn}	Sweep rate of the measurement downsweep.
μ_{meas}	Sweep rate of the sweep for distance measurement.
μ_{ts}	Sweep rate of the transponder.
μ_{up}	Sweep rate of the measurement upsweep.
n	Integer index.
$n_{e,i}$	Number of expected counts in a bin.
$n_{o,i}$	Number of observed counts in a bin.
N	Number of data points used for the FFT or IFFT.
N_{bs}	Number of data points used for the real FFT in the base station.

N_{pll}	PLL N-divider ratio.
N_{ts}	Number of data points used for the real FFT in the transponder.
R_{pll}	PLL R-divider ratio.
RRW	Ramp rate word of the DDS.
σ_f	Standard deviation of the measured frequencies f_{up} and f_{dn}.
σ_n	Standard deviation of a noise process.
t	The time.
$\hat{t}$	Alternative time basis, $\hat{t} = 0$ when the sampling of the low-pass filtered mixed signal starts.
t_d	Time-of-flight of a signal traveling from one station to another, proportional to the distance of the stations.
$t_{d,dn}$	Time-of-flight corresponding to the distance of the stations during the measurement downsweep.
$t_{d_{sync}}$	Time-of-flight corresponding to the distance of the stations during the synchronization upsweep.
$t_{d_{sync,k}}$	$t_{d_{sync}}$ corresponding to the distance of the base station and the transponder in channel k.
$t_{d,up}$	Time-of-flight corresponding to the distance of the stations during the measurement upsweep.
$2t_d$	Round-trip time-of-flight of a signal traveling from one station to another and back.
$2t_{d_{max}}$	Round-trip time-of-flight corresponding to the maximum measurement range of the system d_{max}.
$2t_{d_{max,1}}$	Round-trip time-of-flight corresponding to the maximum distance in measurement channel 1.
$2t_{d_{max,k}}$	Round-trip time-of-flight corresponding to the maximum distance in measurement channel k.
T	Duration of a sweep.
T_1	Duration of the first synchronization sweep.
T_2	Duration of the second synchronization sweep.
T_{bs}	Duration of the sweeps of the base station.
T_c	Duration of a complete measurement cycle.
T_{meas}	Duration of the sweep for distance measurement.
T_p	Time required by the transponder for sampling and evaluation of the low-pass filtered mixed signal during synchronization.
T_t	Nominal time between the trigger events of two consecutive sweeps.

$T_{t,bs}$	Time between the trigger events of two consecutive sweeps in the base station.
T_{ti}	Time between two consecutive ticks of the DSP timer used to trigger the sweeps.
T_{ts}	Duration of the sweeps of the transponder.
$T_{t,ts}$	Time between the trigger events of two consecutive sweeps in the transponder.
δt_1	Estimation error of Δt_1.
δt_{dn}	Estimation error of Δt_{dn}.
δt_{up}	Estimation error of Δt_{up}.
Δt	Offset in time of the local signal in the transponder with respect to the received signal.
$\lvert \Delta t \rvert_{max}$	Maximum allowable offset in time.
Δt_1	Offset in time during the synchronization upsweep.
Δt_2	Offset in time during the synchronization downsweep.
Δt_{dn}	Expected offset in time during the measurement downsweep.
Δt_{dn}^D	Estimate of Δt_{dn} if both stations move relatively to each other.
Δt_{dn}^E	Estimate of Δt_{dn} including estimation errors.
Δt_{laser}	Time between two consecutive laser measurements.
Δt_{lpr}	Time between two consecutive LPR measurements.
Δt_0^M	Offset in time of the component of the received signal corresponding to the LOS path.
Δt_1^M	Offset in time of the component of the received signal corresponding to the first (shortest) NLOS path.
Δt_i^M	Offset in time of the component of the received signal corresponding to NLOS path i.
Δt_I^M	Offset in time of the component of the received signal corresponding to the longest NLOS path I.
Δt_r	Remaining offset in time after synchronization due to the resolution of the DSP timer.
Δt_{up}	Expected offset in time during the measurement upsweep.
Δt_{up}^D	Estimate of Δt_{up} if both stations move relatively to each other.
Δt_{up}^E	Estimate of Δt_{up} including estimation errors.
τ_0	Propagation delay of the radar signal corresponding to the LOS path.
τ_1	Propagation delay of the radar signal corresponding to the first (shortest) NLOS path.
τ_i	Propagation delay of the radar signal corresponding to NLOS path i.

τ_{i+1}	Propagation delay of the radar signal corresponding to NLOS path $(i+1)$.
τ_I	Propagation delay of the radar signal corresponding to the longest NLOS path I.
v	Relative velocity of the radar stations, positive if the stations are moving towards each other.
v_k	Relative velocity of the base station and the transponder in measurement channel k.
v_{max}	Maximum possible relative velocity of the radar units.
v_{meas}	Measured relative velocity.
v_{meas}^A	Measured relative velocity if f_{up} and f_{dn} are aliased.
v_{min}	Minimum possible relative velocity of the radar units.
v_{off}	Offset in the measured relative velocity if f_{up} and f_{dn} are aliased.
$w(n)$	Window coefficients.
x	A signal.
x_{bs}	Low-pass filtered mixed signal in the base station.
$x_{bs,lo}$	Locally generated signal in the base station.
$x_{bs,rx}$	Signal received by the base station.
$x_{bs,rx}^\alpha$	Signal received by the base station when the mismatch of the sweep rates of the radar stations is considered.
$x_{bs,rx}^D$	Signal received by the base station when the Doppler frequency shift is considered.
$x_{bs,rx0}^M$	Component of the signal received by the base station corresponding to the LOS path.
$x_{bs,rx1}^M$	Component of the signal received by the base station corresponding to the first (shortest) NLOS path.
$x_{bs,tx}$	Signal transmitted by the base station.
x_{lo}	Local signal.
x_{rx}	Received signal.
x_{ts}	Low-pass filtered mixed signal in the transponder.
$x_{ts,lo}$	Locally generated signal in the transponder.
$x_{ts,lo}^\alpha$	Locally generated signal in the transponder when the mismatch of the sweep rates of the stations is considered.
$x_{ts,mix}$	Mixed signal in the transponder.
$x_{ts,rx}$	Signal received by the transponder.

$x_{ts,rx}^{D}$	Signal received by the transponder when the Doppler frequency shift is considered.
$x_{ts,rx0}^{M}$	Component of the signal received by the transponder corresponding to the LOS path.
$x_{ts,rx1}^{M}$	Component of the signal received by the transponder corresponding to the first (shortest) NLOS path.
$x_{ts,tx}$	Signal transmitted by the transponder.
x_{tx}	Transmitted signal.
$\underline{X}$	FFT of the signal x.
$\underline{X}_{bs}$	FFT of the low-pass filtered mixed signal x_{bs} in the base station.
$\underline{X}_{bs}^{Z}$	Zoom FFT around the peak magnitude of $\underline{X}_{bs}$.
$\left|\underline{X}_{mz-1}^{Z}\right|$	Magnitude of the zoom FFT spectrum one bin left of the peak magnitude.
$\left|\underline{X}_{mz}^{Z}\right|$	Peak magnitude of the zoom FFT spectrum.
$\left|\underline{X}_{mz+1}^{Z}\right|$	Magnitude of the zoom FFT spectrum one bin right of the peak magnitude.
$\underline{X}_{ts}$	FFT of the low-pass filtered mixed signal x_{ts} in the transponder.
$\hat{\xi}_{n}$	Normalized noise magnitude.
$(\cdot)^{*}$	Complex conjugate of a number.
$\mathcal{FFT}$	Fast Fourier Transform.
$\mathcal{IFFT}$	Inverse Fast Fourier Transform.
$\Longrightarrow$	Logical implication.
∞	Infinity.
$\lvert\cdot\rvert$	Magnitude of a variable.
$\max(\cdot)$	Maximum of a set of values.
$\mathrm{mean}(\cdot)$	Mean of a random variable.
$\min(\cdot)$	Minimum of a set of values.
$\mathrm{pdf}(\cdot)$	Probability density function of a random variable.
$\mathrm{std}(\cdot)$	Standard deviation of a random variable.
$\mathrm{var}(\cdot)$	Variance of a random variable.

Chapter 1 — Introduction

In the last decade, the demand for precise location and tracking of people, vehicles, or objects has increased tremendously. There are many exciting applications with a huge market potential for wireless location solutions. Today, wireless location systems are widely used in both consumer and industrial applications.

Consumer applications include mobile personal navigation systems, location based services, and augmented reality [102]. Personal navigation systems provide interactive maps guiding the user to a selected target. They are commonly used in cars as well as with mobile communication devices like cell phones or Personal Digital Assistants (PDAs). Location based services provide useful information depending on the users current position, e. g. information on traffic jams or personalized weather services [88]. Finally, the emerging field of augmented reality combines real-world and computer-generated data to provide additional information to the users. For instance, computer graphics can be superimposed onto real-world imagery to provide operating instructions for complex systems or visualize hidden objects [41].

Industrial applications range from simple location of vehicles and tools to safety-related applications. The precise tracking of industrial transportation means in real-time, for instance, is a main requirement for the optimization of logistics processes and storage management. If the position of cranes, forklifts, etc. is tracked continuously the position of the transported goods is known as well. If the position of all goods in a storage area is managed in a suitable data base, the goods can be located and picked up reliably at any time. Further industrial applications for location systems include crane collision avoidance and the control of Automated Guided Vehicles (AGVs).

Location systems should provide accurate and consistent position measurements. The required accuracy, however, strongly depends on the application. For consumer applications like car navigation an accuracy of a few meters often is sufficient. Collision avoidance applications typically require an accuracy on the order of ten centimeters, while the automated operation of vehicles calls for accuracies down to the centimeter-range.

Current location systems are either based on ultrasound, optical, or radar principles. Ultrasound ranging systems only provide short range measurements up to a few meters [36]. They suffer from a high propagation loss and are sensitive to environmental conditions [19]. The accuracy of ultrasound location systems is typically on the order of a few decimeters [36, 56].

Laser ranging systems provide accurate measurements over ranges of more than one kilometer with accuracies in the centimeter-range [49]. Their application, however, is limited by environmental conditions as well. Strong background light lowers the achieved accuracy significantly [32]. Furthermore, optical systems are distorted by fog and dirt deposit. Especially for long range distance measurements the laser must be adjusted precisely to a suitable retroreflector.

Radar systems on the other hand offer a higher flexibility than laser systems since they can be equipped with a wide range of antennas. Thus, directional antennas can be used for 1D distance measurements, while omni-directional antennas can provide a full 360° beam width. Furthermore, radar systems are robust to environmental conditions like fog and dirt deposit. Their accuracy strongly depends on the bandwidth of the radar signals. It is typically on the order of a few centimeters to a few decimeters. Since radar systems provide a good accuracy and a high flexibility they are well-suited for a wide range of industrial applications.

Perhaps the most popular location systems are global navigation satellite systems, e. g. the American NAVSTAR Global Positioning System (GPS), the European Galileo system, and the Russian GLObal NAvigation SAtellite System (GLONASS) [17,18,108]. These systems use a set of Earth orbiting satellites as known references that transmit radio signals. The signals allow dedicated receivers to estimate their location and velocity. Satellite based navigation systems, by nature, require a line-of-sight between the receiver and a sufficient number of satellites in order to provide an accurate estimate of the position. Line-of-sight conditions, however, cannot be guaranteed in industrial environments. Clearly, the line-of-sight to the satellites is obstructed in buildings.

This immanent disadvantage of global navigation satellite systems can be overcome by applying local positioning solutions. Here, a set of reference units is placed at known positions in the area of interest. Mobile units then estimate their position relative to the known reference markers. Mainly, three different measurement principles are used for local positioning systems today: Angle-Of-Arrival (AOA), Received-Signal Strength (RSS), and propagation-time based systems [103].

In AOA systems the position is calculated via goniometry. The angle or bearing relative to reference markers located at known positions is measured with the use of directional antennas or antenna arrays. The position value is then given by the intersection of several measured direction pointers. The accuracy of AOA systems, however, is limited by the possible directivity of the measuring aperture, by shadowing, and by multipath reflections arriving from misleading directions [103].

RSS systems are based on propagation-loss equations. The free-space path loss, for instance, is proportional to $1/d^2$, where d is the distance between the units. However, this simple equation is in most cases unsuited to calculate the distance value from the difference of the transmitted and the received power under real conditions [50,103]. In indoor environments or built-up areas, multipath fading and shadowing have a dominant effect [22,103]. To overcome this problem RSS-based location fingerprinting can be used [51]. Here, the actual field distribution in the area of interest is learned from measurements first. The recorded field distribution is then compared to the real-time measurement results during the actual location process [40,103].

Propagation-time based location systems, in general, provide more precise position estimates than AOA or RSS systems. They are divided into Time-Of-Arrival (TOA), Round-trip Time-Of-Flight (RTOF), and Time-Difference-Of-Arrival (TDOA) systems [101].

In TOA systems the distance between the stations is calculated from the one-way propagation time of a signal traveling from the transmitter to the receiver. The TOA concept requires precise synchronization of all stations involved [28]. For example, a timing synchronization below 30 ps is required for a positioning accuracy in the centimeter-range. Since the clock information has to be distributed to and kept in all units this approach either leads to a very expensive or less accurate system [103].

For RTOF systems the clock synchronization requirement can somewhat be relaxed. Here, the second station (transponder) replies to the signal received from the first station. The distance between the stations is then calculated from the complete round-trip propagation time of a signal. To achieve high precision measurements the exact processing time of the transponder must be known to the first station. If the transponder merely reflects the signal received from the first unit this synchronization problem is avoided [45, 96]. Passive reflectors, however, only allow for a short measurement range since the level of the radar signals then decays with $1/d^4$. Active reflectors are usually sensitive to clutter noise from the environment. In this thesis, however, an RTOF system is presented, where the transponder synchronizes to signals from the first station. After a well-known delay it then generates a synchronized reply which is transmitted back to the first station. The RTOF system topology described in section 1.2.3, therefore, allows for robust long range measurements.

The third group of propagation-time based positioning systems used today are TDOA systems, where the time-difference of signals arriving at multiple measuring units is used to calculate the position of a mobile unit. TDOA systems, however, suffer from conditioning problems since the position of the objects to be tracked is given by the intersection of hyperbolas. They have been shown to perform worse or, at best, strictly equal to TOA location systems, where the position of the objects is given by the intersection of circles [98]. In TDOA systems the measuring units have to be synchronized precisely. In practice, the synchronization is done by using a dedicated reference transponder in a known position [90] or by a backbone network between the measuring units [53]. The RTOF system presented in this thesis, on the other hand, does not require any additional infrastructure. Due to the simple and yet powerful system architecture it is well-suited for a wide range of applications.

Regardless of the measurement principle applied, RSS, TOA, RTOF, or TDOA, the key requirement for precise position estimates are accurate 1D distance measurements between the mobile unit and the reference stations. For 2D and 3D applications the 1D distance measurements to multiple reference units have to be combined appropriately [23, 24, 34]. The accuracy of the position estimates of the wireless location systems can further be improved by incorporating data from inertial sensors [54, 91]. The focus of the work at hand, however, is on precise 1D distance measurements.

1.1 Outline of the Thesis

In this thesis a novel high performance RTOF radar system for precise distance and velocity measurement is presented and analyzed. Two radar stations, a base station and a transponder, measure their distance and relative velocity similar to the Frequency-

Modulated Continuous Wave (FMCW) radar principle. Therefore, the remainder of chapter 1 is dedicated to a description of the FMCW radar principle. The basic topology of the distance and velocity measurement system is introduced as well.

Chapter 2 provides a thorough mathematical analysis of the underlying algorithm for distance measurement. The analysis is significantly expanded over the analysis in previous publications [69]. For the first time the effect of relative motion of the radar stations is taken into account for the derivation of the equations. It is shown how the relative velocity of the radar stations can be estimated from the Doppler frequency shift of the radar signals. Furthermore, novel concepts to multiplex the signals of multiple transponders are developed and analyzed, thus enabling the base station to measure its distance to multiple transponder units which is a prerequisite for 2D positioning. Finally, it is shown how the measurement range of the system, which is currently limited by the computational power of the signal processor, can be doubled without increasing the computational complexity if the velocity measurements are evaluated carefully.

The algorithms at hand are used to implement a real-life measurement system. The design of the system realized in this thesis is described in chapter 3. First, an overview over the actual implementation of the measurement cycle is given. Requirements for the signal generator are then derived from the algorithms and the configuration of the current hardware setup is described. Finally, relevant system parameters like the center frequency and the bandwidth of the radar signals are given. These parameters provide the basis for the simulations and measurements throughout the remainder of the thesis.

Special emphasis is placed on a detailed system-theoretic analysis of the measurement process. In chapter 4 possible sources of error are identified, and their effect on the accuracy of the distance measurements is investigated. For instance, the influence of multipath propagation, the Signal-to-Noise Ratio (SNR), and the stability of the clock frequencies of the radar stations are studied in detail. Furthermore, the algorithm for frequency estimation is presented. Its theoretical inaccuracy is obtained by simulation as well.

Finally, the hardware setup presented in chapter 3 is tested in various environments. The results of an extensive measurement campaign are presented in chapter 5. The test scenarios include free-space measurements on an airfield as well as measurements in an indoor environment. For some measurements the results of the radar system presented here are compared to the results of an industrial laser ranging system. The results show the excellent performance of the algorithms and the hardware setup presented in this thesis.

1.2 Concept of the Novel Radar System

The algorithm for distance and velocity measurement analyzed in this thesis is closely related to the well-known FMCW radar principle. Therefore, relevant parameters of FMCW radar signals are described in the following. A short introduction to the FMCW radar principle is given as well. Finally, the important modifications to the FMCW radar principle are highlighted, and the topology of the distance and velocity measurement system is presented.

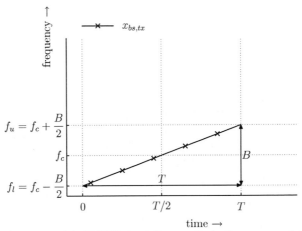

Figure 1.1: Frequency of an FMCW signal (base station): Frequency-modulated continuous wave signals are used for synchronization and measurement. At time $t = 0$ the base station transmits a signal with linear frequency modulation. Here, the frequency increases linearly with time.

1.2.1 Frequency-Modulated Continuous Wave Signals

The algorithms described in the following chapter utilize FMCW radar signals with linear frequency modulation to synchronize two radar stations and to subsequently measure their distance and relative velocity. In general, the frequency of the signal is swept from a lower frequency limit f_l to an upper frequency limit f_u or vice versa. The sweep is called an upsweep if the frequency of the signal is increased during the sweep. The term downsweep is used if the frequency is decreased. The instantaneous frequency of an FMCW signal during an upsweep is depicted in figure 1.1.

In this thesis each sweep is characterized by the sweep duration T, the center frequency f_c, and the sweep bandwidth B which is given by:

$$B = f_u - f_l. \tag{1.1}$$

The ratio of the sweep bandwidth and the sweep duration is referred to as the sweep rate, which is described by:

$$\mu = \frac{B}{T}. \tag{1.2}$$

Note that the lower frequency limit f_l of the sweeps can be used instead of the center frequency f_c to describe the FMCW radar signals. This has been done by the author in previous publications [69, 70]. The solution to the synchronization and distance measurement problem, however, is more intuitive if the center frequency is used to derive the equations. This especially holds if different sweep rates in the base station and the transponder are considered.

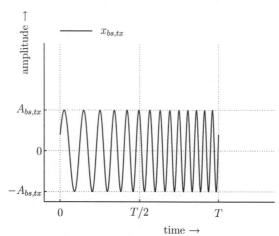

Figure 1.2: Amplitude of an FMCW signal (base station): The amplitude of the FMCW signal is not changed during transmission, the frequency is modulated only.

Figure 1.2 depicts the signal corresponding to the upsweep shown in figure 1.1. The amplitude $A_{bs,tx}$ of the signal does not change during the sweep, only the frequency is modulated.

1.2.2 Frequency-Modulated Continuous Wave Radar Principle

A radar signal with linear frequency modulation as shown in figure 1.1 can be utilized to measure the distance between a radar unit and a target. The FMCW radar principle is well-known [44, 52, 92]. The basic concept is introduced briefly in the following.

Figure 1.3 depicts the working principle of a primary radar system. The radar station transmits an FMCW radar signal. The signal is reflected by the passive target after the time-of-flight t_d. The time t_d required by the radar signal to propagate from the radar unit to the target is proportional to the distance between the radar station and the target.

The time required by the reflected signal to propagate back to the radar station is again t_d. The reflected signal, therefore, arrives at the radar station after the RTOF $2t_d$. Clearly, the distance between the radar unit and the target can be calculated if the RTOF is measured and the phase velocity of the radar signal is known.

Figure 1.4 depicts the instantaneous frequency of the signals $x_{bs,tx}$ and $x_{bs,rx}$ which are transmitted and received by the radar station at time $t = 0$ and $t = 2t_d$, respectively. Note that the RTOF $2t_d$ usually is much smaller than the duration T of the radar signals. A large $2t_d$ has been chosen in figure 1.4 for clarity of presentation only.

The transmitted signal $x_{bs,tx}$ and the received signal $x_{bs,rx}$ are multiplied in a mixer and the resulting signal is low-pass filtered. The frequency f_d of the low-pass filtered mixed signal is then proportional to the RTOF $2t_d$. After the frequency of the low-pass

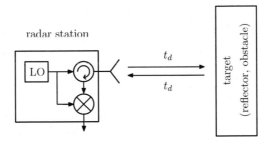

Figure 1.3: Setup of a primary radar system: A radar station transmits a signal with linear frequency modulation which is generated by the local oscillator (LO). The signal is reflected by an obstacle. The reflected signal arrives at the radar station after the round-trip time-of-flight $2t_d$. If the phase velocity of the radar signal is known the distance from the radar station to the obstacle can be calculated from the round-trip time-of-flight.

filtered mixed signal is estimated, the RTOF and the distance between the radar station and the target can be calculated subsequently.

1.2.3 System Topology

The common FMCW radar principle described in the previous section is modified for the system presented in this thesis. The modified principle is patented [82]. In the following the modifications to the common FMCW radar principle are described briefly.

Firstly, a secondary radar system is used. Here, the passive target in figure 1.3 is replaced by an active radar unit. The use of active targets enables a larger range of the radar system since a signal is actively transmitted back from the target to the first radar station rather than only being reflected. Furthermore, multiple targets can be separated and identified if each target uses a unique modulation for the signal transmitted back to the first unit [45, 107].

Figure 1.5 shows the setup of the basic 1D measurement system. It consists of two radar stations. Both units share an identical hardware setup, which is described in detail in chapter 3. A detailed block diagram of the radar units is depicted in figure 3.2. Each station, however, is configured in software to work in one of two modes. The first unit is set to work as a base station (BS). It measures its distance and relative velocity to a second unit working in transponder mode (TS).

Again, the RTOF $2t_d$ of a signal traveling from the base station to the transponder and back is used to estimate the distance between both radar stations. The base station, therefore, transmits a signal as described in section 1.2.1 to the transponder. The transponder then synchronizes to the impinging signal with high precision. A synchronized reply is sent back to the base station after a precisely known processing

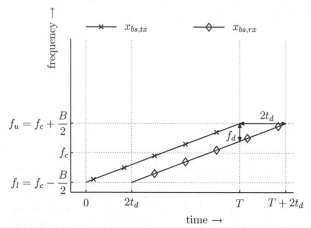

Figure 1.4: FMCW radar principle (base station): The signal $x_{bs,tx}$ is transmitted by the
radar station at time $t = 0$. It is reflected by an obstacle. The reflected signal
$x_{bs,rx}$ arrives at the radar station after the round-trip time-of-flight $2t_d$, which is
proportional to the distance of the stations. If $x_{bs,tx}$ and $x_{bs,rx}$ are multiplied, the
frequency of the low-pass filtered mixed signal f_d is proportional to t_d. Thus, the
distance between the stations can be calculated from the frequency of the low-pass
filtered mixed signal.

delay T_p. The processing delay is a constant known to both radar units. The delay is
required for the calculations in the transponder during synchronization.

When the synchronized reply arrives at the base station, its delay $(T_p + 2t_d)$ depends
linearly on the RTOF of the signal. Since T_p is known to the base station the RTOF
$2t_d$ and thus the distance d of both units can be calculated. If the Doppler frequency
shift f_D of the synchronized reply is evaluated the relative velocity of the radar stations
can be measured as well.

A detailed derivation of the algorithms for synchronization and measurement is given
in the next chapter.

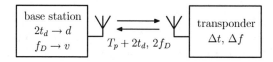

Figure 1.5: Setup of the secondary radar system: The base station transmits a signal to the
transponder. The transponder synchronizes to the impinging signal and sends a
synchronized reply to the base station. There, the distance and the relative velocity
of both units are calculated from the delay $2t_d$ and the Doppler frequency shift f_D
of the reply. A detailed block diagram of the radar units is depicted in figure 3.2.

Chapter 2 — Distance and Velocity Measurement Principle

In this chapter the concept of the FMCW radar system for precise distance and velocity measurements, which has been introduced in Section 1.2.3, is derived in detail. The system consists of two radar units. The first unit is configured as a base station. The base station measures its distance and relative velocity to a second unit, which is referred to as the transponder. The measurement process is comprised of two important steps, the synchronization of the radar units and the actual distance and velocity measurement. During synchronization the base station transmits an FMCW radar signal to the transponder. The transponder synchronizes to the impinging signal with high precision. After the synchronization is completed successfully the transponder transmits a synchronized FMCW radar signal back to the base station. In the base station the distance and relative velocity of both stations are then calculated from the RTOF and the Doppler frequency shift of the received reply, respectively.

Chapter 2 is organized as follows. In section 2.1 it is shown how FMCW radar signals can be utilized to synchronize two radar stations with high precision. Subsequently, the algorithm to measure the distance of both radar stations is analyzed as well. For the initial analysis it is assumed that the radar units do not move relatively to each other during the measurement and the Doppler frequency shift can be neglected.

However, if the stations move relatively to each other the Doppler frequency shift of the radar signals has to be taken into account. Therefore, a novel extension of the algorithms for synchronization and distance measurement is introduced in section 2.4. A detailed mathematical analysis of the synchronization and measurement process shows how the Doppler frequency shift can be exploited to measure the relative velocity of the radar stations in addition to their distance.

A 2D location system can be set up if the distances from the base station to multiple transponders are measured. Various multiplexing schemes for the signals of multiple transponders are analyzed in section 2.5. A very efficient multiplexing scheme similar to the well-known Frequency Division Multiple Access (FDMA) principle is introduced within this thesis, where the distances to all transponders are measured with a single measurement sweep.

Finally, the maximum range of the hardware setup presented in chapter 3 is addressed. It is currently limited by the computational power of the signal processor and by the sampling frequency used to digitize the signals. However, a modified version of the algorithm for distance and velocity measurement is derived in section 2.7. The novel algorithm allows to extend the range of the measurement system without increasing the computational complexity.

The derivations of the algorithms for synchronization as well as for distance and velocity measurement provided in this chapter are very detailed. Therefore, the final results are summarized in section 2.2 and section 2.6, respectively.

2.1 Synchronization of the Stations

The measurement system at hand utilizes the round-trip time-of-flight of a signal traveling from the base station to the transponder and back to determine the distance between the two stations. Both stations have to be synchronized with high precision before the transponder can send a synchronized signal back to the base station. In the following, a general solution to the synchronization problem is developed. The general solution, however, is then simplified for practical application.

2.1.1 Synchronization Problem

The synchronization starts when the base station transmits a signal $x_{bs,tx}$ with linear frequency modulation at time $t = 0$. The signal $x_{bs,tx}$ transmitted by the base station has already been shown in figure 1.2. Its instantaneous frequency during the interval $(0 \leq t \leq T)$ is given by:

$$f_{bs,tx} = f_c + \mu \left(t - \frac{T}{2} \right), \tag{2.1}$$

where the sweep rate μ depends on the bandwidth and the duration of the sweep (1.2). The instantaneous frequency of the transmitted signal is shown in figure 1.1.

From (2.1) the phase of the signal $x_{bs,tx}$ transmitted by the base station is obtained by integrating the angular frequency $2\pi f_{bs,tx}$ over time. Hence, the signal $x_{bs,tx}$ is represented by:

$$x_{bs,tx} = A_{bs,tx} \cos \left(2\pi f_c t + \pi \mu t^2 - \pi \mu T t + \varphi_0 \right), \tag{2.2}$$

where $A_{bs,tx}$ is the amplitude of the signal transmitted by the base station and φ_0 is an arbitrary phase term resulting from the integration of (2.1).

The signal $x_{bs,tx}$ transmitted by the base station arrives at the transponder at time $t = t_d$. The time-of-flight t_d is related to the distance d between both units by:

$$t_d = \frac{d}{c_{ph}}, \tag{2.3}$$

where c_{ph} is the phase velocity of the radar signals. The phase velocity in air depends on atmospheric conditions [15, 64]. Usually it is slightly smaller than the speed of light in vacuum.

If line-of-sight propagation is assumed, the signal received by the transponder is a delayed and attenuated copy of the transmitted signal. The received signal $x_{ts,rx}$ and its phase $\varphi_{ts,rx}$ are given by:

$$x_{ts,rx} = A_{ts,rx} \cos \left(2\pi f_c \left(t - t_d \right) + \pi \mu \left(t - t_d \right)^2 - \pi \mu T \left(t - t_d \right) + \varphi_1 \right), \tag{2.4}$$

$$\varphi_{ts,rx} = 2\pi f_c \left(t - t_d \right) + \pi \mu \left(t - t_d \right)^2 - \pi \mu T \left(t - t_d \right) + \varphi_1, \tag{2.5}$$

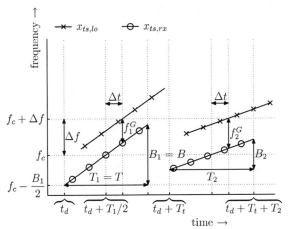

Figure 2.1: Synchronization, general solution (transponder): At time $t = t_d$ the signal transmitted by the base station is received by the transponder. The received and the locally generated signal are multiplied in a mixer. Then, the offset in time Δt and the offset in frequency Δf of both signals are calculated from the frequencies f_1^G and f_2^G of the low-pass filtered mixed signal during two consecutive sweeps of different sweep rate.

respectively, where $A_{ts,rx}$ is the amplitude of the received signal and φ_1 is an arbitrary constant phase term. Equations (2.4) and (2.5) are valid for $(t_d \leq t \leq t_d + T)$. The instantaneous frequency of the signal $x_{ts,rx}$ received by the transponder is depicted in figure 2.1.

The transponder generates a local signal which is given by:

$$x_{ts,lo} = A_{ts,lo} \cos\left(2\pi\left(f_c + \Delta f\right)\left(t - t_d - \Delta t\right) + \pi\alpha\mu\left(t - t_d - \Delta t\right)^2\right.$$
$$\left. - \pi\alpha\mu T\left(t - t_d - \Delta t\right) + \varphi_2\right). \tag{2.6}$$

Here, $A_{ts,lo}$ is the amplitude of the local signal in the transponder and φ_2 is another arbitrary constant phase term. The phase of the locally generated signal is given by:

$$\varphi_{ts,lo} = 2\pi\left(f_c + \Delta f\right)\left(t - t_d - \Delta t\right) + \pi\alpha\mu\left(t - t_d - \Delta t\right)^2$$
$$-\pi\alpha\mu T\left(t - t_d - \Delta t\right) + \varphi_2. \tag{2.7}$$

Equations (2.6) and (2.7) are valid for $(t_d + \Delta t \leq t \leq t_d + \Delta t + T)$. The instantaneous frequency of the local signal $x_{ts,lo}$, described by (2.6), is depicted in figure 2.1 as well.

The local signal in the transponder $x_{ts,lo}$ is similar to the received signal $x_{ts,rx}$. The output frequency of the signal generator of each station, however, is proportional to the clock frequency of the station. In general, the clock frequencies of the base station $f_{clk,bs}$ and the clock frequency of the transponder $f_{clk,ts}$ do not match exactly. The frequencies of the clock oscillators differ slightly from their nominal values due to manufacturing tolerances, temperature drift or aging [60, 80]. Therefore, the center frequency of the locally generated signal differs from the center frequency f_c of the received signal by an offset in frequency Δf, which is assumed to be positive if the center frequency of the local signal $x_{ts,lo}$ is larger than the center frequency of the received signal $x_{ts,rx}$, i. e.:

$$(f_{clk,ts} > f_{clk,bs}) \implies (f_c + \Delta f > f_c). \qquad (2.8)$$

The offset in frequency is negative if the center frequency of the local signal in the transponder is smaller than the center frequency of the received signal, i. e.:

$$(f_{clk,ts} < f_{clk,bs}) \implies (f_c + \Delta f < f_c). \qquad (2.9)$$

In practice the offset in frequency is much smaller than the bandwidth of the radar signals. A large offset in frequency, however, is used in figure 2.1 for clarity of presentation.

Furthermore, the locally generated signal is delayed by an offset in time Δt with respect to the received signal. The offset in time is caused by random activation of each of the modules and the variations of the clock sources of the stations. It is observed at the center frequencies of the sweeps of both stations and is assumed to be positive if the received signal $x_{ts,rx}$ precedes the local signal $x_{ts,lo}$ in the transponder. The offset in time is negative if the local signal in the transponder precedes the received signal.

Finally, the sweep rate of the locally generated signal differs from the sweep rate μ of the received signal by a factor of α, which depends on the difference of the clock frequencies of the base station and the transponder. It is larger than 1 if the clock frequency of the transponder is larger than the clock frequency of the base station. Similarly, α is less than 1 if the clock frequency of the base station exceeds the clock frequency of the transponder, i. e.:

$$(f_{clk,ts} > f_{clk,bs}) \implies (\alpha > 1), \qquad (2.10)$$
$$(f_{clk,ts} < f_{clk,bs}) \implies (\alpha < 1). \qquad (2.11)$$

Table 2.1 lists relevant parameters of the received signal and the local signal in the transponder prior to synchronization. For now only the first sweep depicted in figure 2.1 is analyzed. The second sweep is discussed subsequently.

2.1.2 General Solution

During synchronization the offsets in time Δt and in frequency Δf are estimated and corrected. Thus, the instantaneous frequency of the local signal $x_{ts,lo}$ in the transponder after synchronization matches the instantaneous frequency of the received signal $x_{ts,rx}$. A synchronized reply then can be sent back to the base station.

To estimate the offsets in time and in frequency the local signal $x_{ts,lo}$ and the received signal $x_{ts,rx}$ are multiplied in a mixer. The mixed signal is given by:

$$x_{ts,mix} = x_{ts,lo}\, x_{ts,rx} \qquad (2.12)$$
$$= A_{ts,lo}\, A_{ts,rx} \cos\left(\varphi_{ts,lo}\right) \cos\left(\varphi_{ts,rx}\right), \qquad (2.13)$$

Table 2.1: Signal parameters prior to synchronization (transponder): The center frequencies, sweep rates and starting points in time of the received signal $x_{ts,rx}$ and the locally generated signal $x_{ts,lo}$ in the transponder differ from each other prior to synchronization.

signal	first sweep		second sweep	
	$x_{ts,rx}$	$x_{ts,lo}$	$x_{ts,rx}$	$x_{ts,lo}$
center frequency	f_c	$f_c + \Delta f$	f_c	$f_c + \Delta f$
sweep rate	$\mu_1 = \mu = \dfrac{B}{T}$	$\alpha\mu$	$\mu_2 = \dfrac{B_2}{T}$	$\alpha\mu_2$
delay with respect to $t = 0$	t_d	$t_d + \Delta t$	$t_d + T_t$	$t_d + T_t + \Delta t$

where $\varphi_{ts,lo}$ and $\varphi_{ts,rx}$ denote the phase of the locally generated signal and the received signal in the transponder according to (2.7) and (2.5).

The product of trigonometric functions in (2.13) can be expressed by a sum of two sinusoidal signals [87]:

$$x_{ts,mix} = \frac{A_{ts,lo} \, A_{ts,rx}}{2} \big(\cos \left(\varphi_{ts,lo} + \varphi_{ts,rx} \right) + \cos \left(\varphi_{ts,lo} - \varphi_{ts,rx} \right) \big). \tag{2.14}$$

The first term of the sum in (2.14) is a signal with a high frequency of approximately twice the center frequency of the sweeps $2f_c$. The second term is a baseband signal with a frequency well below f_c.

Hence, the mixed signal is low-pass filtered and only the second term remains:

$$x_{ts} = \frac{A_{ts,lo} \, A_{ts,rx}}{2} \cos \left(\varphi_{ts,lo} - \varphi_{ts,rx} \right). \tag{2.15}$$

The phase of the low-pass filtered mixed signal is given by:

$$\varphi_{ts} = \varphi_{ts,lo} - \varphi_{ts,rx} \tag{2.16}$$

$$= 2\pi t \Delta f - 2\pi \alpha\mu \left(t - t_d \right) \Delta t + \pi\mu \left(t - t_d \right)^2 \left(\alpha - 1 \right)$$
$$- \pi\mu T \left(t - t_d \right) \left(\alpha - 1 \right) + \varphi_3, \tag{2.17}$$

where the components of $\varphi_{ts,lo}$ and $\varphi_{ts,rx}$ that do not depend on the time t are summed up to the constant phase term φ_3.

The instantaneous frequency of the low-pass filtered mixed signal f_{ts} is obtained by differentiating its phase with respect to time t:

$$f_{ts} = \frac{1}{2\pi} \frac{\mathrm{d}}{\mathrm{dt}} \left(\varphi_{ts} \right) \tag{2.18}$$

$$= \Delta f - \alpha\mu\Delta t + \mu \left(t - t_d - \frac{T}{2} \right) \left(\alpha - 1 \right). \tag{2.19}$$

Equation (2.19) is valid during the interval:

$$\max(t_d, t_d + \Delta t) \leq t \leq \min(t_d + T, t_d + T + \Delta t), \tag{2.20}$$

when the received signal (2.4) and the local signal (2.6) in the transponder are both defined. The operators $\min(\cdot)$ and $\max(\cdot)$ are required, since the offset in time Δt can be positive or negative.

An equation similar to (2.19) can be derived for the second synchronization sweep depicted in figure 2.1. The second sweep is transmitted by the base station after the trigger time T_t. Consequently, the corresponding signal arrives at the transponder at time $T_t + t_d$. During the second sweep the instantaneous frequency of the signal received by the transponder is given by:

$$f_{ts,rx} = f_c + \mu_2 \left(t - T_t - t_d - \frac{T_2}{2} \right) \qquad (2.21)$$

for $(T_t + t_d \leq t \leq T_t + t_d + T_2)$. Here, T_2 denotes the duration of the second sweep. The instantaneous frequency of the locally generated signal is then described by:

$$f_{ts,lo} = f_c + \Delta f + \alpha \mu_2 \left(t - T_t - t_d - \Delta t - \frac{T_2}{2} \right) \qquad (2.22)$$

for $(T_t + t_d + \Delta t \leq t \leq T_t + t_d + \Delta t + T_2)$.

The instantaneous frequency of the low-pass filtered mixed signal in the transponder during the second sweep is expressed as the difference of the instantaneous frequencies of both signals, i. e.:

$$
\begin{aligned}
f_{ts} &= f_{ts,lo} - f_{ts,rx} & (2.23) \\
&= \Delta f - \alpha \mu_2 \Delta t + \mu_2 \left(t - T_t - t_d - \frac{T_2}{2} \right) (\alpha - 1). & (2.24)
\end{aligned}
$$

Equation (2.24) is valid during the interval:

$$\max(t_d, t_d + \Delta t) \leq (t - T_t) \leq \min(t_d + T_2, t_d + T_2 + \Delta t), \qquad (2.25)$$

when the equations for the instantaneous frequencies of the received signal (2.21) and the local signal (2.22) in the transponder are both valid. Again, the operators $\min(\cdot)$ and $\max(\cdot)$ are required, since the offset in time Δt can be positive or negative.

The frequency of the low-pass filtered mixed signal during the first and second synchronization sweep is given by (2.19) and (2.24), respectively. Here, the time is referenced absolutely to the instant $t = 0$ where the first synchronization sweep has been transmitted by the base station.

On the other hand the transponder has no knowledge of this time instant. It estimates the frequency of its low-pass filtered mixed signal by sampling the signal and applying a Fast Fourier Transform (FFT) algorithm. For the first synchronization sweep sampling is started synchronously to the local sweep at time $t = t_d + \Delta t$. The sampling of the second synchronization sweep is again started synchronously to the corresponding local sweep at time $t = t_d + \Delta t + T_t$. After all samples are acquired in the transponder the FFT is applied to the digitized signal and the position of the peak magnitude in the power spectral densities is detected. Then, the frequency of the low-pass filtered mixed signal is calculated from the peak position.

Note, that a separate FFT is calculated for each synchronization sweep. Naturally, the FFT results in the transponder are referenced relatively to the time the sampling for each synchronization sweep is started. Therefore, a common time basis must be established for both synchronization sweeps. This can be achieved by replacing the absolute time t in (2.19) and (2.24) by an alternative time basis $\hat{t}$. The alternative time basis $\hat{t}$ is chosen such that $\hat{t} = 0$ at the time the sampling for each sweep is started, i. e.:

$$\hat{t} = \begin{cases} t - t_d - \Delta t & \text{for the first synchronization sweep} \\ t - t_d - \Delta t - T_t & \text{for the second synchronization sweep} \end{cases} \quad (2.26)$$

If (2.26) is applied to (2.19) and (2.24) the frequencies of the low-pass filtered mixed signal f_1^G and f_2^G obtained from the spectral analysis during the first and second synchronization sweep, respectively, can be expressed with respect to the common time basis $\hat{t}$ as:

$$f_1^G = \Delta f - \alpha\mu_1\Delta t + \mu_1\left(\hat{t} + \Delta t - \frac{T_1}{2}\right)(\alpha - 1), \quad (2.27)$$

$$f_2^G = \Delta f - \alpha\mu_2\Delta t + \mu_2\left(\hat{t} + \Delta t - \frac{T_2}{2}\right)(\alpha - 1). \quad (2.28)$$

In (2.27) $T_1 = T$ denotes the duration of the first synchronization sweep.

In general, the frequencies f_1^G and f_2^G are not constant over time. Equation (2.27) is interpreted as a chirp around a constant frequency $(\Delta f - \alpha\mu_1\Delta t)$. Similarly, (2.28) describes a chirp around a constant frequency $(\Delta f - \alpha\mu_2\Delta t)$. The term chirp is used to avoid confusion with the sweeps of the FMCW radar signals. The bandwidth of the chirps depends on the ratio α of the sweep rates of both radar stations as well as on the sweep rate, and thus on the bandwidth, of the radar signals. Therefore, the more α differs from $\alpha = 1$, the larger the chirp bandwidth becomes. This is treated in detail in section 4.7.

The linear set of equations given by (2.27) and (2.28) is solved with respect to Δt and Δf. The offset in time Δt and the offset in frequency Δf are then given by:

$$\Delta t = \frac{f_2^G - f_1^G}{(\mu_1 - \mu_2)} + (\alpha - 1)\left(\hat{t} - \frac{\mu_1 T_1 - \mu_2 T_2}{2(\mu_1 - \mu_2)}\right), \quad (2.29)$$

$$\Delta f = \frac{\mu_1 f_2^G - \mu_2 f_1^G}{\mu_1 - \mu_2} - (\alpha - 1)\frac{\mu_1\mu_2}{\mu_1 - \mu_2}\frac{T_1 - T_2}{2}. \quad (2.30)$$

In general, they cannot be calculated exactly if the sweep rates of both stations do not match $(\alpha \neq 1)$. However, it can be shown that the ratio of the sweep rates α is linked to the offset in frequency Δf.

Deviation of the Clock Frequencies and Mismatch of the Sweep Rates

The relative deviation of the clock frequencies of the base station and the transponder is given by:

$$\delta_{clk} = \frac{f_{clk,ts} - f_{clk,bs}}{f_{clk,bs}}, \quad (2.31)$$

where $f_{clk,bs}$ and $f_{clk,ts}$ denote the clock frequencies of the base station and the transponder, respectively.

For the hardware setup presented in chapter 3, the output frequency of the signal generator of each station is proportional to its clock frequency. Therefore, the ratio of the center frequencies of the sweeps generated by both stations matches the ratio of their clock frequencies as well as the ratio of the bandwidths of their sweeps, i. e.:

$$\frac{f_c + \Delta f}{f_c} = \frac{f_{clk,ts}}{f_{clk,bs}} = \frac{B_{ts}}{B_{bs}} = 1 + \delta_{clk}. \tag{2.32}$$

Consequently, the sweeps of the station with the higher clock frequency have a higher bandwidth and center frequency.

Since the internal time in timer-ticks "passes faster" on the station with the higher clock frequency, the ratio of the duration of the sweeps in both stations is the inverse of the ratio of their clock frequencies, i. e.:

$$\frac{T_{ts}}{T_{bs}} = \frac{f_{clk,bs}}{f_{clk,ts}} = \frac{1}{1 + \delta_{clk}}. \tag{2.33}$$

Thus, the ratio of the sweep rates of the transponder μ_{ts} and the base station μ_{bs} can be calculated directly from the offset in frequency Δf by applying (2.32) and (2.33) to the definition of α. The ratio is given by:

$$\alpha = \frac{\mu_{ts}}{\mu_{bs}} = \frac{B_{ts}}{B_{bs}} \frac{T_{bs}}{T_{ts}} \tag{2.34}$$

$$= (1 + \delta_{clk})^2 = \left(1 + \frac{\Delta f}{f_c}\right)^2. \tag{2.35}$$

Hence, the ratio of the sweep rates of the base station and the transponder is known as well if the offset in frequency has been determined.

2.1.3 Simplification of the Equations

If the parameters of the signals during both synchronization sweeps are chosen appropriately, (2.27) to (2.30) can be simplified. In the following, this is discussed for two reasonable choices of sweep parameters for practical application.

Sweeps of Equal Bandwidth and Duration, but Opposite Sweep Direction

A straightforward solution to the synchronization problem is obtained if the duration of the first synchronization sweep matches the duration of the second synchronization sweep. Two sweeps of equal duration are transmitted, i. e.:

$$T_1 = T_2 = T. \tag{2.36}$$

Consequently, (2.30) simplifies to:

$$\Delta f = \frac{\mu_1 f_2^G - \mu_2 f_1^G}{\mu_1 - \mu_2}, \tag{2.37}$$

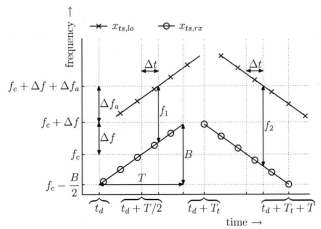

Figure 2.2: Synchronization, equal bandwidth (transponder): If sweeps of equal bandwidth and opposite sweep rate, i. e. $B_1 = B$, $T_1 = T$, $B_2 = -B$, $T_2 = T$, are used during synchronization, the equations for Δt and Δf can be simplified. Furthermore, an additional offset in frequency Δf_a is added to the local signal to ensure a positive frequency of the low-pass filtered mixed signal.

and the offset in frequency is estimated correctly even if $\alpha \neq 1$. Subsequently, the deviation δ_{clk} of the clock frequencies of both stations and the ratio α of the sweep rates of the local signal and the received signal in the transponder are calculated from (2.32) and (2.35). Consequently, the sweep rate of the transponder can be adjusted for a second synchronization step. It then matches the sweep rate of the base station. Therefore, it is assumed in the following that the sweep rates of the radar stations match:

$$\alpha = 1. \tag{2.38}$$

Furthermore, it is assumed that the bandwidth B of the first synchronization sweep matches the bandwidth of the second synchronization sweep. The sweep direction, however, is changed from the first to the second sweep as depicted in figure 2.2, i. e. an upsweep and a downsweep of equal bandwidth are used for synchronization.

The sweep rates of the synchronization upsweep and downsweep are then given by:

$$\mu_1 = -\mu_2 = \mu. \tag{2.39}$$

Consequently, (2.27) and (2.28) simplify to:

$$f_1 = \Delta f - \mu\Delta t, \tag{2.40}$$
$$f_2 = \Delta f + \mu\Delta t, \tag{2.41}$$

and the constant frequencies f_1 and f_2 of the low-pass filtered mixed signal in the transponder during the synchronization upsweep and downsweep do not depend on $\hat{t}$ anymore.

The frequencies f_1 and f_2 according to (2.40) and (2.41) are positive or negative, depending on the offset in time Δt and the offset in frequency Δf prior to synchronization. Therefore, a complex valued FFT is required to determine the magnitude and sign of both frequencies. For practical application, however, it is preferable to use real valued samples of the low-pass filtered mixed signal and a real valued FFT. Then, only the magnitude of the frequencies f_1 and f_2 is obtained from the power spectral densities and the sign cannot be determined.

To overcome this problem an additional offset in frequency Δf_a is introduced [69, 70]. It is a constant system parameter and is added to the frequency of the local sweeps in the transponder $x_{ts,lo}$ during synchronization as depicted in figure 2.2. If the condition

$$\Delta f_a \geq |\Delta f|_{max} + \mu |\Delta t|_{max} \tag{2.42}$$

is satisfied, the frequency of the low-pass filtered mixed signal x_{ts} is always positive [75] and a real valued FFT can be used to estimate the frequency of the low-pass filtered mixed signal. Here, $|\Delta t|_{max}$ is the maximum allowable offset in time and $|\Delta f|_{max}$ is the maximum allowable offset in frequency prior to synchronization. Both $|\Delta t|_{max}$ and $|\Delta f|_{max}$ depend on the actual hardware implementation of the synchronization principle described here. An example for choosing Δf_a is given in section 4.3.1.

If Δf_a is added to the local signal of the transponder Δf has to be substituted by $(\Delta f + \Delta f_a)$ in (2.40) and (2.41), i. e.:

$$f_1 = \Delta f + \Delta f_a - \mu \Delta t, \tag{2.43}$$

$$f_2 = \Delta f + \Delta f_a + \mu \Delta t. \tag{2.44}$$

The frequencies f_1 and f_2 then depend solely on the offset in time Δt, the offset in frequency Δf, and known system parameters.

Therefore, the linear set of equations given by (2.43) and (2.44) is solved with respect to the offset in time and the offset in frequency:

$$\Delta t = \frac{f_2 - f_1}{2\mu}, \tag{2.45}$$

$$\Delta f = \frac{f_2 + f_1}{2} - \Delta f_a, \tag{2.46}$$

where the frequencies f_1 and f_2 are estimated from the samples of the low-pass filtered mixed signal using the FFT algorithm. After Δt and Δf are calculated the transponder adjusts its local signal. The instantaneous frequency of the local FMCW radar signal in the transponder $x_{ts,lo}$ then matches the instantaneous frequency of the received signal $x_{ts,rx}$ and a synchronized reply can be sent back to the base station.

Mono-Frequent Signal and a Single Sweep

Another choice of sweep parameters is shown in figure 2.3. Again, an additional offset in frequency Δf_a is added to the frequency of the locally generated signal in the transponder. The bandwidth and hence the sweep rate of the first sweep is set to:

$$B_1 = 0 \quad \Longrightarrow \quad \mu_1 = 0. \tag{2.47}$$

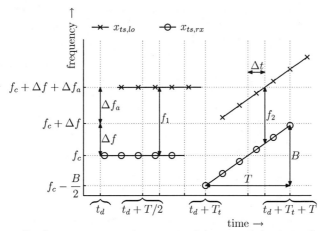

Figure 2.3: Synchronization, mono-frequent signal (transponder): A signal with a constant frequency, e. g. f_c, can be used to measure the offset in frequency Δf directly, i. e. $B_1 = 0$. Then a single sweep suffices to estimate the offset in time Δt.

Consequently, the first synchronization sweep is a signal with a constant frequency [69]. The offset in frequency then is obtained directly from (2.27):

$$\Delta f = f_1 - \Delta f_a. \tag{2.48}$$

Hence, the deviation δ_{clk} of the clock frequencies of the transponder and the base station and the ratio α of the sweep rates of the local signal and the received signal can be calculated from (2.32) and (2.35), respectively. The sweep rate of transponder then can be adjusted and (2.38) holds.

The signal with a constant frequency, i. e. the first sweep with bandwidth $B_1 = 0$, is followed by a second sweep as shown in figure 2.3. The sweep rate of the second sweep is:

$$\mu_2 = \mu = \frac{B_2}{T_2} = \frac{B}{T}. \tag{2.49}$$

The offset in time is calculated subsequently from (2.28):

$$\Delta t = \frac{\Delta f + \Delta f_a - f_2}{\mu}. \tag{2.50}$$

This approach is picked up by Gierlich et al. to synchronize their radar stations [26]. They use a signal with a constant frequency to measure the deviation of the clock frequencies of two stations. Subsequently, a single sweep is used to estimate the offset in time of a received and a local signal.

The system presented in this thesis, however, uses an upsweep and a downsweep of equal bandwidth and duration (2.39) for synchronization. Therefore, the theory on synchronization presented in the remainder of this thesis is based on (2.43) and (2.44). The advantages of this approach are shown subsequently when the relative motion of the radar stations is included in the calculations.

However, for now it is still assumed that the radar stations do not move relatively to each other. In practice, two problems remain to be solved to precisely synchronize the base station and the transponder. Firstly, the offset in time Δt is not constant during synchronization if the clock frequencies of both stations do not match. Secondly, the radar stations have to be pre-synchronized due to technical limitations of the measurement hardware. Both issues are addressed in the following.

2.1.4 Time Dependent Offset in Time

Until now it has been assumed that the offset in time Δt between the received signal and the local signal in the transponder does not change from the synchronization upsweep to the synchronization downsweep. However, the synchronization downsweep is triggered T_t after the synchronization upsweep is started. The interval T_t is defined by a certain number of ticks of a timer in each station where the number of ticks is identical in both modules. The timer of each station, however, is clocked with a frequency proportional to the clock frequency of the station. Hence, the time corresponding to a constant number of timer-ticks is inversely proportional to the clock frequency of the timer.

Consequently, if the clock frequencies of both stations do not match T_t "passes faster" on one of the stations, i. e.:

$$\frac{f_{clk,ts}}{f_{clk,bs}} = \frac{f_c + \Delta f}{f_c} = \frac{T_{t,bs}}{T_{t,ts}} = 1 + \delta_{clk}, \tag{2.51}$$

and the offset in time during the synchronization upsweep differs from the offset in time during the synchronization downsweep. Therefore, (2.43) to (2.46) have to be extended for high precision synchronization.

The offset in time during the synchronization downsweep Δt_2 is given by the offset in time during the synchronization upsweep Δt_1 and the slightly different trigger intervals in the transponder $T_{t,ts}$ and in the base station $T_{t,bs}$:

$$\Delta t_2 = \Delta t_1 + (T_{t,ts} - T_{t,bs}). \tag{2.52}$$

Now $T_{t,ts}$ is arbitrarily chosen as the correct value of T_t:

$$T_t = T_{t,ts}, \tag{2.53}$$

and the offset in time during the synchronization downsweep is calculated using (2.51):

$$\Delta t_2 = \Delta t_1 + T_t \left(1 - \frac{T_{t,bs}}{T_{t,ts}} \right) \tag{2.54}$$

$$= \Delta t_1 - T_t \frac{\Delta f}{f_c}. \tag{2.55}$$

In practice, the deviation δ_{clk} of the clock frequencies of the radar stations is typically on the order of 10^{-5} or smaller. Therefore, the same result is obtained from (2.52) if $T_{t,bs}$ is chosen as the true value of T_t and the approximation

$$\frac{1}{1 + \delta_{clk}} \approx 1 - \delta_{clk} \tag{2.56}$$

is used [12].

Consequently, (2.43) and (2.44) are rewritten as:

$$f_1 = \Delta f + \Delta f_a - \mu \Delta t_1, \tag{2.57}$$

$$f_2 = \Delta f + \Delta f_a + \mu \left(\Delta t_1 - T_t \frac{\Delta f}{f_c} \right), \tag{2.58}$$

and the linear set of equations given by (2.57) and (2.58) is solved with respect to the offset in frequency and the offset in time during the synchronization upsweep:

$$\Delta f = \frac{f_2 + f_1 - 2\Delta f_a}{2 - \dfrac{T_t}{T}\dfrac{B}{f_c}}, \tag{2.59}$$

$$\Delta t_1 = \frac{f_2 - f_1}{2\mu} + \frac{T_t}{2}\frac{\Delta f}{f_c}. \tag{2.60}$$

Equations (2.59) and (2.60) can be used to calculate and correct the offset in frequency and the offset in time of the local signal in the transponder with respect to the received signal during the synchronization upsweep with high precision. It has already been stated, however, that a synchronized reply is sent back from the transponder to the base station to measure the distance of the radar stations. In section 2.3.2 it will be shown that the synchronized reply is comprised of an upsweep and a downsweep of equal bandwidth. The first measurement sweep is transmitted T_p after the synchronization upsweep, for the second measurement sweep the delay is $(T_p + T_t)$. The processing delay T_p is chosen such that the transponder can finish the synchronization process before transmitting the measurement sweeps.

Similarly to (2.51) and (2.52), both processing delays depend on the clock frequencies of the base station and the transponder. Therefore, the processing delays in both stations do not match exactly and the offset in time changes during synchronization and measurement. Thus, the offsets in time which need to be corrected for the measurement upsweep and downsweep are given by:

$$\Delta t_{up} = \Delta t_1 - T_p \frac{\Delta f}{f_c}, \tag{2.61}$$

$$\Delta t_{dn} = \Delta t_1 - (T_p + T_t) \frac{\Delta f}{f_c}. \tag{2.62}$$

The measurement sweeps of the base station and the transponder are then synchronized with high precision.

2.1.5 Pre-Synchronization of the Stations

The frequencies of the low-pass filtered mixed signal x_{ts} in the transponder during the synchronization upsweep f_1 and downsweep f_2 are given by (2.43) and (2.44). It is preferable to use real valued samples of the low-pass filtered mixed signal and a real valued FFT to estimate f_1 and f_2 to keep the hardware design as simple as possible. Therefore, a positive frequency of the low-pass filtered mixed signal in the transponder must be guaranteed for all possible combinations of Δf and Δt, i. e.:

$$f_{1,min} > 0. \tag{2.63}$$

In section 2.1.3 it has been shown that this can be achieved by an additional offset in frequency Δf_a which is added to the frequency of the local sweeps in the transponder. The required minimum additional offset in frequency is given by (2.42).

From (2.42) and (2.43) the maximum possible frequency of the low-pass filtered mixed signal then can be derived. It is given by:

$$f_{1,max} \geq |\Delta f|_{max} + \Delta f_a + \mu |\Delta t|_{max} \tag{2.64}$$

$$\geq 2 \left(|\Delta f|_{max} + \mu |\Delta t|_{max} \right). \tag{2.65}$$

Consequently, the minimum sampling frequency the signal x_{ts} has to be digitized with to avoid aliasing is [84]:

$$f_s \geq 2 f_{1,max} \tag{2.66}$$

$$\geq 4 \left(|\Delta f|_{max} + \frac{B}{T} |\Delta t|_{max} \right). \tag{2.67}$$

In (2.67) $|\Delta t|_{max}$ is the maximum allowable offset in time and $|\Delta f|_{max}$ is the maximum allowable offset in frequency prior to synchronization. The latter is a constant system parameter which is determined by the stability of the oscillators used to clock the radar stations and the center frequency of the FMCW radar signals. The sweep duration T is a constant system parameter as well.

On the other hand the sampling frequency f_s is a constant given by the hardware used to digitize the low-pass filtered mixed signal. In practice, the sampling frequency f_s usually is much smaller than the full bandwidth $B_f = B$ of the synchronization sweeps. The maximum frequency of the low-pass filtered mixed signal $f_{1,max}$, therefore, must be reduced to avoid aliasing. From (2.67) this is achieved by either decreasing the bandwidth B of the synchronization sweeps or by reducing the maximum offset in time $|\Delta t|_{max}$. Both methods of pre-synchronization are discussed next.

Adaptive Control of the Bandwidth of the Synchronization Sweeps

If the initial maximum offset in time $|\Delta t|_{max}$ is large, e. g. $|\Delta t|_{max} > 0.5T$, the bandwidth of the synchronization sweeps can be increased gradually to achieve high precision synchronization of the radar stations. The number of required synchronization steps, i. e. sweep bandwidths, depends on the sampling frequency of the transponder and the final bandwidth B_f of the synchronization sweeps. After each synchronization step

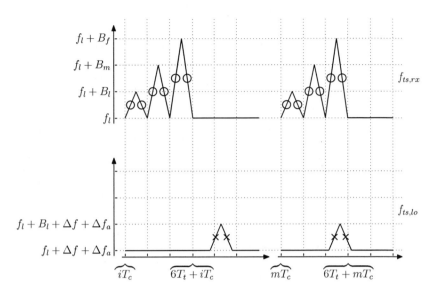

Figure 2.4: Pre-synchronization, low bandwidth (transponder): The base station transmits a cyclic pattern of synchronization sweeps with bandwidth B_l, B_m, and B_f. Initially, the transponder generates local sweeps with a low bandwidth B_l. As long as the sweeps with low bandwidth of both stations do not overlap no peaks are found in the power spectral densities of the low-pass filtered mixed signal in the transponder.

the remaining maximum offset in time $|\Delta t|_{max}$ is reduced and the bandwidth can be increased for the next step. In the following, an example is given where three different bandwidths are utilized to synchronize the transponder to the base station.

The synchronization sweeps are transmitted by the base station in a cycle pattern. First, an upsweep and a downsweep with low bandwidth B_l are transmitted. They are followed by two sweeps with medium bandwidth B_m as well as an upsweep and a downsweep with the full bandwidth B_f of the synchronization sweeps. This pattern is repeated after the cycle time T_c. The cycle time is chosen such that the transponder has ample time to finish the calculations required for the synchronization that have been described in the previous sections.

The signal transmitted by the base station is received by the transponder. The instantaneous frequency $f_{ts,rx}$ of the sweeps received by the transponder is depicted in the first row of figure 2.4 to figure 2.6. The time-of-flight of the radar signals has been omitted in all three figures since it is much smaller than the sweep duration in practice.

Initially, the transponder is set to generate local sweeps of bandwidth B_l. The frequency $f_{ts,lo}$ of the local sweeps with low bandwidth is depicted in the second row of figure 2.4. In each measurement cycle the locally generated signal is mixed with the

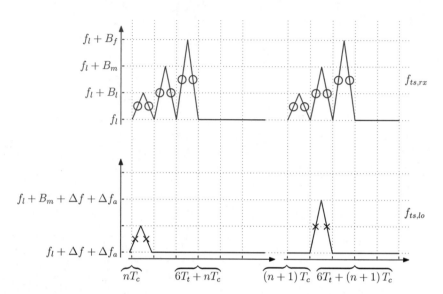

Figure 2.5: Pre-synchronization, medium bandwidth (transponder): Eventually, the sweeps
with bandwidth B_l of both stations overlap. Then, peaks at the frequencies f_1 and
f_2 are detected in the power spectral densities of the low-pass filtered mixed signal
and the offset in time is corrected. The synchronization bandwidth of the local
signal in the transponder can then be increased to B_m and the synchronization
cycle is repeated. Thus, a more accurate estimate of the offset in time is obtained.

received signal during the local upsweep and downsweep in the transponder. The low-
pass filtered mixed signal is digitized and analyzed using the synchronization algorithm
described in the previous sections.

On the left-hand side of figure 2.4 the cyclic pattern of the synchronization sweeps has
been repeated i-times already. The local sweeps in the transponder do not overlap with
the low-bandwidth sweeps of the base station. Therefore, no peaks will be detected in
the power spectral density of the low-pass filtered mixed signal. The transponder then
adjusts its local timer by a fraction of the time between consecutive sweeps, e. g. $T_t/10$,
and the synchronization cycle is repeated. On the right-hand side of figure 2.4 the
synchronization cycle has been repeated m-times. The local sweeps in the transponder
still do not overlap with the corresponding sweeps of the received signal. Therefore, the
local timer in the transponder has to be adjusted again.

Eventually, the time bases of the base station and the transponder match coarsely.
The synchronization sweeps with low bandwidth of the received and the local signal
in the transponder then overlap for a significant fraction of the sweep duration. This
is shown on the left-hand side of figure 2.5 where the synchronization cycle has been
repeated n-times.

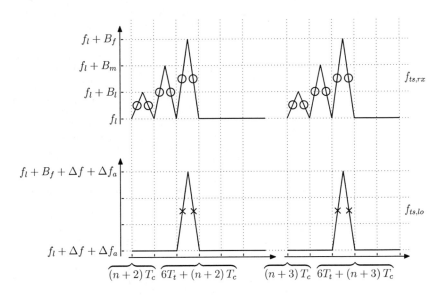

Figure 2.6: Pre-synchronization, high bandwidth (transponder): After both stations are synchronized at the medium bandwidth B_m the remaining offset in time is small compared to the sweep duration. Thus, the full synchronization bandwidth B_f can finally be utilized to synchronize the radar stations with high precision.

Since the sweeps of the received and the local signal overlap, peaks at the frequencies f_1 and f_2 are detected in the power spectral densities of the low-pass filtered mixed signal during the upsweep and the downsweep, respectively. Substituting B by B_l, (2.45) is used to estimate and correct the offset in time for the first time.

During the next synchronization cycle the offset in time of the received signal and the locally generated signal in the transponder is much smaller. Therefore, the bandwidth of the synchronization sweeps is increased to B_m. This is shown on the right-hand side of figure 2.5. Again the low-pass filtered mixed signal in the transponder is sampled during the synchronization sweeps and the frequencies f_1 and f_2 are obtained from the spectral analysis. Thus, a more accurate estimate of the offset in time is obtained from (2.45), substituting B by B_m, and the transponder is synchronized to the base station more precisely.

Finally, synchronization sweeps with the full bandwidth $B_f = B$ are used for the synchronization process as shown in figure 2.6. Again, the low-pass filtered mixed signal is sampled during the local sweeps in the transponder. The frequencies f_1 and f_2 obtained from the spectral analysis match the frequencies given by (2.57) and (2.58). Now, the radar stations can be synchronized with high precision if (2.59) and (2.60) are used to estimate the offset in time Δt and the offset in frequency Δf. The distance measurement can easily be included in the cyclic pattern described here as well [73].

The pre-synchronization principle described above allows to synchronize the transponder to the base station even if the initial offset in time is large. There is no additional hardware required to implement the algorithm. The number of pre-synchronization steps, i. e. the number of different sweep bandwidths, is chosen depending on the full synchronization bandwidth and the initial maximum offsets in time and in frequency. The time passing until the sweeps with low bandwidth overlap and the first successful synchronization is achieved, however, can be lengthy since it depends on the initial offset in time. To overcome this problem a communication based pre-synchronization concept is used. This approach is presented next.

Limitation of the Initial Offset in Time by Utilizing a Communication Channel

If the full bandwidth $B_f = B$ is used for the first synchronization sweeps the offset in time prior to synchronization must be small compared to the sweep duration T (2.67). The exact value of the maximum allowable offset in time $|\Delta t|_{max}$ depends on the sampling frequency used to digitize the low-pass filtered mixed signal in the transponder, the bandwidth B of the sweeps, and the maximum offset in frequency prior to synchronization $|\Delta f|_{max}$. In the following, a communication scheme is discussed that allows to synchronize the transponder to the base station with an inaccuracy significantly below $1\,\mu s$.

The concept is ideally suited for the measurement system setup presented in the next chapter. It utilizes Frequency Shift Keying (FSK) to pre-synchronize the radar stations. The signal generator used for generating the synchronization sweeps in the base station and the transponder can also be used to generate the FSK signals. Only minimal hardware overhead is required to receive the FSK modulated signals. Since similar frequencies are used for the FSK signals and the synchronization sweeps the communication range approximately matches the measurement range of the system. Furthermore, the same antennas can be used to transmit and receive the FSK and the FMCW radar signals.

The concept for pre-synchronization is shown in figure 2.7. The base station transmits a data sequence, e. g. a sequence of thirty 0-1 sequences. The frequency of the signal transmitted by the base station is toggled between two frequencies $f_{fsk,0}$ and $f_{fsk,1}$ corresponding to transmitting the symbols 0 or 1, respectively.

The signal is received by the transponder and multiplied with a local signal with constant frequency $f_{fsk,lo}$. The mixed signal is band-pass filtered and demodulated. Edges in the received data sequence are detected and a local timer in the transponder is adjusted to the edges of the received sequence with an inaccuracy well below $0.4\,\mu s$. Finally, after a specific communication sequence, e. g. three 000-111 sequences, is received the synchronization process described in the previous sections is started. It will be shown in section 4.3.1 that an initial offset in time below $0.4\,\mu s$ is sufficient to start the synchronization with the full bandwidth of the hardware setup presented in section 3.

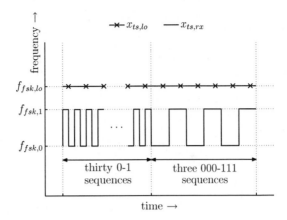

Figure 2.7: Pre-synchronization, communication channel (transponder): The base station transmits a data sequence using frequency shift keying. In the transponder, the received signal and the locally generated signal are multiplied. The mixed signal is band-pass filtered and fed to an FSK receiver chip. The digital output signal is evaluated. Edges in the received data sequence, e. g. thirty 0-1 sequences, are detected and a local timer is adjusted to the edges. Finally, the synchronization process starts if a specific data sequence, e. g. three 000-111 sequences, is received.

Note, that it is mandatory to transmit the pre-synchronization sequence from the base station to the transponder. If it was transmitted by the transponder and received by the base station the time bases of both units could be pre-synchronized as well. However, in this case an error would remain after pre-synchronization. The error depends linearly on the propagation time of the radar signals and hence on the distance of the stations. Pre-synchronization, therefore, would become more inaccurate as the distance between the stations increases. Thus, it is mandatory to synchronize the transponder to an FSK signal transmitted by the base station.

2.2 Summary of the Algorithm for Synchronization

In the previous sections an algorithm for the precise synchronization of two radar stations has been presented. It can be summarized as follows.

The synchronization process is initiated when the base station transmits an FSK signal. The FSK sequence is received by the transponder. There, a local timer is synchronized to the edges of the received data sequence. Consequently, the time basis of the transponder coarsely matches the time basis of the base station after the FSK sequence has been received.

The base station then transmits an upsweep and a downsweep, i. e. a signal with linear frequency modulation. Both synchronization sweeps are received by the transponder.

The transponder itself generates a local signal similar to the received signal. The locally generated signal in the transponder, however, differs from the received signal by an offset in time Δt and an offset in frequency Δf.

To estimate both offsets the local signal and the received signal are multiplied in a mixer. The resulting signal is low-pass filtered and digitized. Subsequently, an FFT algorithm is used to determine the frequencies of the low-pass filtered mixed signal during the synchronization upsweep f_1 and the synchronization downsweep f_2.

The offset in frequency of both stations is then given by (2.59) which is repeated here for convenience:

$$\Delta f = \frac{f_2 + f_1 - 2\Delta f_a}{2 - \dfrac{T_t}{T}\dfrac{B}{f_c}}. \tag{2.59}$$

The offsets in time that need to be corrected for the measurement upsweep and downsweep are calculated from the frequencies f_1 and f_2 as well. They are given by:

$$\Delta t_{up} = \frac{T}{B}\frac{f_2 - f_1}{2} + \frac{T_t}{2}\frac{\Delta f}{f_c} - T_p\frac{\Delta f}{f_c}, \tag{2.68}$$

$$\Delta t_{dn} = \frac{T}{B}\frac{f_2 - f_1}{2} - \frac{T_t}{2}\frac{\Delta f}{f_c} - T_p\frac{\Delta f}{f_c}. \tag{2.69}$$

The bandwidth B, the center frequency f_c, the duration T of the synchronization sweeps, and the additional offset in frequency Δf_a are known system parameters. The time T_t between two consecutive sweeps and the processing delay T_p between synchronization and measurement are known constants as well. Equations (2.68) and (2.69) are obtained by applying (2.60) to (2.61) and (2.62), respectively.

2.3 Distance Measurement

After the two radar stations are synchronized their distance can be measured. The distance measurement principle is described subsequently in this section. The principle is similar to the well-known FMCW radar approach [44, 52, 92]. In the following, the general solution is derived assuming different sweep rates of the sweeps of the base station and the transponder. For the derivation it is also assumed that the offsets in frequency and in time were estimated correctly in the transponder. The general solution is simplified subsequently in section 2.3.2. Finally, the effect of synchronization errors is investigated.

2.3.1 General Solution

The distance measurement principle is closely related to the synchronization principle described before. The distance measurement starts after the processing delay T_p with respect to the instant $t = 0$, when the first synchronization sweep in figure 2.1 had been transmitted by the base station. Again the base station generates a local sweep of bandwidth B_{meas} and duration T_{meas}. This time, however, the sweep is not transmitted.

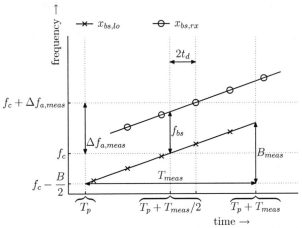

Figure 2.8: Distance measurement, general solution (base station): After both stations are synchronized, the transponder sends a synchronized reply back to the base station. The synchronized reply is received by the base station at time $t = (T_p+2t_d)$. Again, the received signal is multiplied with a locally generated signal in a mixer. The frequency of the low-pass filtered mixed signal depends linearly on the round-trip time-of-flight $2t_d$ of the signal. Therefore, the distance of both stations is calculated from the frequency f_{bs} of the low-pass filtered mixed signal during the sweep.

The instantaneous frequency of the local sweep $x_{bs,lo}$ in the base station is depicted in figure 2.8. The sweep is described by:

$$x_{bs,lo} = f_c + \mu_{meas}\left(t - T_p - \frac{T_{meas}}{2}\right) \qquad (2.70)$$

for $(T_p \le t \le T_p + T_{meas})$. In (2.70) f_c is the center frequency of the sweeps and T_p is a known processing delay used by the transponder to synchronize its signal. The sweep rate of the measurement sweep is:

$$\mu_{meas} = \frac{B_{meas}}{T_{meas}}. \qquad (2.71)$$

During synchronization the local signal of the transponder was synchronized relatively to the time $t = t_d$ at which the synchronization sweeps were received by the transponder (2.59), (2.68). The transponder transmits its synchronized reply T_p after the synchronization started. Thus, the total delay with respect to $t = 0$ of the synchronized reply transmitted by the transponder is $(t_d + T_p)$. The instantaneous frequency of the signal transmitted by the transponder, therefore, is given by:

$$f_{ts,tx} = f_c + \Delta f_{a,meas} + \alpha \mu_{meas}\left(t - T_p - t_d - \frac{T_{meas}}{2}\right), \qquad (2.72)$$

where an additional offset in frequency $\Delta f_{a,meas}$ has been added to the sweep transmitted by the transponder. Again, the sweep rate of the sweep generated by the transponder differs from the sweep rate of the sweep generated by the base station by a factor α due to the deviation of the clock frequencies of both stations.

The synchronized reply of the transponder arrives at the base station after the time-of-flight t_d corresponding to the distance of the stations. The instantaneous frequency of the signal $x_{bs,rx}$ received by the base station is given by:

$$f_{bs,rx} = f_c + \Delta f_{a,meas} + \alpha\mu_{meas}\left(t - T_p - 2t_d - \frac{T_{meas}}{2}\right) \qquad (2.73)$$

for $(T_p + 2t_d \leq t \leq T_p + 2t_d + T_{meas})$. It is also depicted in figure 2.8.

The received signal $x_{bs,rx}$ and the locally generated signal $x_{bs,lo}$ are multiplied in a mixer in the base station. The mixed signal is low-pass filtered and digitized. Similarly to (2.24), the instantaneous frequency of the low-pass filtered mixed signal in the base station is given by the difference of the instantaneous frequencies of the received and the local signal:

$$\begin{aligned} f_{bs} &= f_{bs,rx} - f_{bs,lo} & (2.74) \\ &= \Delta f_{a,meas} - \alpha\mu_{meas}(2t_d) + \mu_{meas}\left(t - T_p - \frac{T_{meas}}{2}\right)(\alpha - 1). & (2.75) \end{aligned}$$

Equation (2.75) is valid for:

$$T_p + 2t_d \leq t \leq T_p + T_{meas}. \qquad (2.76)$$

Again, the instant when the sampling of the low-pass filtered mixed signal is started has to be taken into account. For the measurement sweep depicted in figure 2.8 sampling in the base station starts synchronously to the local sweeps at $t = T_p$. Consequently, the time basis in (2.75) and (2.76) is substituted by:

$$\hat{t} = t - T_p, \qquad (2.77)$$

and the general solution for the frequency of the low-pass filtered mixed signal in the base station during distance measurement, given as:

$$f_{bs}^G = \Delta f_{a,meas} - \alpha\mu_{meas}(2t_d) + \mu_{meas}\left(\hat{t} - \frac{T_{meas}}{2}\right)(\alpha - 1), \qquad (2.78)$$

is obtained for $(2t_d \leq \hat{t} \leq T_{meas})$.

Equation (2.78) is similar to (2.27) and (2.28). It is interpreted as a chirp with a low bandwidth around a constant frequency $(\Delta f_{a,meas} - 2t_d\alpha\mu_{meas})$. The bandwidth of the chirp and hence the width of the peak in the power spectral density depends on the ratio of the sweep rates of both stations, i.e. on the factor α, and on the sweep rate μ_{meas} of the measurement sweeps. The more α differs from $\alpha = 1$, the wider the peaks become. The widening of the peaks is treated in detail in section 4.7.

The frequency f_{bs}^G of the low-pass filtered mixed signal is obtained from its power spectral density and the distance is calculated subsequently by solving (2.78) with respect to the time-of-flight t_d. The distance is given by:

$$d = c_{ph}t_d \tag{2.79}$$

$$= \frac{c_{ph}}{2\alpha\mu_{meas}}\left(\Delta f_{a,meas} - f_{bs}^G\right) + \frac{\alpha - 1}{2\alpha}\left(\hat{t} - \frac{T_{meas}}{2}\right). \tag{2.80}$$

The system parameters c_{ph}, B_{meas}, and T_{meas} are known constants. The additional offset in frequency $\Delta f_{a,meas}$ is used to ensure a positive frequency of the low-pass filtered mixed signal. It is a known constant as well. However, if the sweep rates of both stations do not match ($\alpha \neq 1$) an error remains when the distance is estimated, since α has been calculated in the transponder but is unknown to the base station.

2.3.2 Simplification of the Equations and Introduction of a Second Measurement Sweep

Equations (2.78) and (2.80) can be simplified if the sweep rates of the base station and the transponder match. Under the aforementioned assumption $\alpha = 1$ (2.38), the frequency of the low-pass filtered mixed signal in the base station is independent of the time $\hat{t}$. It is given by:

$$f_{bs} = \Delta f_{a,meas} - 2t_d\mu_{meas}. \tag{2.81}$$

This is similar to the well-known FMCW radar distance measurement principle where the reflection of a signal transmitted by a radar station is returned to the station after the round-trip time-of-flight $2t_d$ [52]. The only difference is the additional offset in frequency $\Delta f_{a,meas}$ added to the synchronized reply of the transponder.

Although the distance of the base station and the transponder can be estimated by evaluating a single sweep only (2.80), an upsweep and a downsweep are used for distance measurement as depicted in figure 2.9. By averaging over both sweeps the standard deviation of the measurement results is decreased. More importantly, the relative velocity of both stations can be measured if the Doppler frequency shift of both sweeps is evaluated. Furthermore, measurement errors can be detected if the difference of the distances measured during the upsweep and the downsweep is too large.

In general, any two sweeps with different sweep rates can be used to measure the distance and relative velocity of the radar stations. However, in section 2.4.2 it is shown that a convenient solution is obtained if an upsweep and a downsweep of equal bandwidth are used for the measurements.

The local measurement upsweep in the base station starts at time T_p. Its bandwidth is $B_{meas} = B$. The same bandwidth is used for the measurement downsweep which is generated after time $(T_p + T_t)$. The duration T is constant for both sweeps. Consequently, the sweep rates of both sweeps are given by:

$$\mu_{up} = -\mu_{dn} = \mu = \frac{B}{T}, \tag{2.82}$$

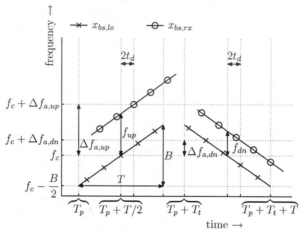

Figure 2.9: Distance measurement, two sweeps (base station): At time $t = T_p + 2t_d$ the measurement upsweep of the transponder arrives at the base station. It is accompanied by a second sweep of negative sweep rate. The received signal and the locally generated signal are multiplied in a mixer. The frequency of the low-pass filtered mixed signal is linearly dependent on the round-trip time-of-flight $2t_d$ of the signal. Thus, the distance of both stations is calculated from the frequency of the low-pass filtered mixed signal during the upsweep f_{up} and the downsweep f_{dn}.

and the frequency of the low-pass filtered mixed signal in the base station during the measurement upsweep and downsweep obtained from (2.81) is:

$$f_{up} = \Delta f_{a,up} - 2t_d\mu, \tag{2.83}$$
$$f_{dn} = \Delta f_{a,dn} + 2t_d\mu. \tag{2.84}$$

If the frequencies f_{up} and f_{dn} are measured, the distance from the base station to the transponder can be calculated from either (2.83) or (2.84) as:

$$d = c_{ph}t_d \tag{2.85}$$
$$= \frac{c_{ph}}{2\mu}\left(\Delta f_{a,up} - f_{up}\right) \tag{2.86}$$
$$= \frac{c_{ph}}{2\mu}\left(f_{dn} - \Delta f_{a,dn}\right). \tag{2.87}$$

Both solutions, however, can be merged to a single equation. The distance between the radar units is then given by:

$$d = \frac{c_{ph}}{4\mu}\left(\left(f_{dn} - \Delta f_{a,dn}\right) - \left(f_{up} - \Delta f_{a,up}\right)\right), \tag{2.88}$$

where c_{ph}, μ, $\Delta f_{a,dn}$, and $\Delta f_{a,up}$ are known system parameters.

2.3.3 Additional Offsets in Frequency and Maximum Measurement Range

The frequency of the low-pass filtered mixed signal in the base station is estimated using the FFT algorithm. To keep the hardware design as simple as possible, it is preferable to use real valued samples of the low-pass filtered mixed signal and a real valued FFT. Therefore, a positive frequency of the low-pass filtered mixed signal must be guaranteed, since positive and negative frequencies cannot be distinguished by the real valued FFT.

To ensure a positive frequency of the low-pass filtered mixed signal in the base station the additional offsets in frequency $\Delta f_{a,up}$ and $\Delta f_{a,dn}$ are added to the synchronized response of the transponder during the measurement upsweep and downsweep, respectively. They are similar to the additional offset in frequency Δf_a that has been added to the local sweeps in the transponder during synchronization.

From (2.83) the frequency of the low-pass filtered mixed signal in the base station during the measurement upsweep f_{up} decreases as the round-trip time-of-flight $2t_d$, i. e. the distance of the radar stations, increases. The minimum frequency is obtained at the maximum measurement range d_{max}. To guarantee a positive frequency of the low-pass filtered mixed signal a large offset in frequency $\Delta f_{a,up}$ is used during the measurement upsweep. It is given by:

$$\Delta f_{a,up} \geq 2\mu \frac{d_{max}}{c_{ph}}. \tag{2.89}$$

Clearly, $\Delta f_{a,up}$ must be larger than the frequency component of the low-pass filtered mixed signal corresponding to the maximum measurement range of the system d_{max}.

The maximum frequency of the low-pass filtered mixed signal in the base station during the measurement upsweep is obtained from (2.83) for $t_d = 0$, i. e. $d = 0$. Clearly, the frequency f_{up} must not exceed the Nyquist frequency [9, 62]. To avoid aliasing during the measurement upsweep the additional offset in frequency is chosen such that:

$$\Delta f_{a,up} \leq \frac{f_s}{2}. \tag{2.90}$$

The frequency of the low-pass filtered mixed signal in the base station during the downsweep f_{dn} increases as the distance increases. Consequently, the minimum frequency is obtained from (2.84) for $t_d = 0$ and only a small additional offset in frequency is required during the measurement downsweep to ensure a positive frequency of the low-pass filtered mixed signal. Therefore, $\Delta f_{a,dn}$ is chosen such that:

$$\Delta f_{a,dn} \geq 0. \tag{2.91}$$

A limit to the measurement range of the system can be derived by applying (2.90) to (2.89). In theory, the maximum measurement range depends on the sampling frequency of the hardware implementation of the algorithms presented here. It is given by:

$$d_{max}^{+} = \frac{c_{ph}}{4\mu} f_s. \tag{2.92}$$

In practice, however, frequencies of the low-pass filtered mixed signal near direct current (DC) or $f_s/2$ have to be avoided, since the results of the spectral analysis are less accurate near the edges of the power spectral densities. This will be shown in section 4.2. Therefore, the additional offset in frequency during the measurement downsweep $\Delta f_{a,dn}$ must be chosen larger than 0 and the additional offset in frequency during the measurement upsweep must be chosen smaller than $f_s/2$. The frequency of the low-pass filtered mixed signal in the base station is then bounded by $\Delta f_{a,dn}$ and $\Delta f_{a,up}$ during both measurement sweeps and the maximum range of the system in practice is given by:

$$d_{max} = \frac{c_{ph}}{2\mu} \left(\Delta f_{a,up} - \Delta f_{a,dn} \right), \qquad (2.93)$$

where the additional offsets $\Delta f_{a,up}$ and $\Delta f_{a,dn}$ are chosen such that:

$$0 < \Delta f_{a,dn} < \Delta f_{a,up} < \frac{f_s}{2}. \qquad (2.94)$$

The theoretical maximum range of the system d_{max}^+, given by (2.92), is an upper limit of d_{max} obtained by setting $\Delta f_{a,up} = f_s/2$ and $\Delta f_{a,dn} = 0$ in (2.93).

2.3.4 Effect of Synchronization Errors

If the frequencies f_1 and f_2 of the low-pass filtered mixed signal during the synchronization sweeps are measured incorrectly in the transponder, the offset in frequency is estimated with an error. This is denoted by:

$$\Delta f^E = \Delta f + \delta f, \qquad (2.95)$$

where Δf^E is the estimated offset in frequency according to (2.59) used to synchronize the local signal in the transponder to the signal of the base station. The true offset in frequency Δf of the radar stations is estimated with an error of δf.

In section 2.1.1 the offset in frequency Δf is assumed to be positive if the center frequency of the sweeps of the transponder is larger than the center frequency of the sweeps of the base station during synchronization. If the offset in frequency is estimated too large, i. e.:

$$\delta f > 0, \qquad (2.96)$$

the measurement upsweep and downsweep transmitted by the transponder start at a frequency that is too low by δf. Consequently, the frequency of the signal $x_{bs,rx}$ received by the base station is decreased by the estimation error. From figure 2.9 it is apparent, that the frequencies of the low-pass filtered mixed signal during the measurement upsweep and downsweep, f_{up} and f_{dn}, are decreased by the estimation error δf as well.

Similarly, the offsets in time for the measurement upsweep and downsweep are estimated with an error. This can be written as:

$$\Delta t_{up}^E = \Delta t_{up} + \delta t_{up}, \qquad (2.97)$$
$$\Delta t_{dn}^E = \Delta t_{dn} + \delta t_{dn}, \qquad (2.98)$$

where Δt_{up}^E and Δt_{dn}^E are the estimated values according to (2.68) and (2.69), respectively. The true offsets in time Δt_{up} and Δt_{dn} need to be corrected to synchronize the measurement sweeps of both stations. They are estimated with an error of δt_{up} and δt_{dn}, respectively.

In section 2.1.1 the offset in time Δt is assumed to be positive if the received signal in the transponder $x_{ts,rx}$ precedes the local signal $x_{ts,lo}$ during synchronization. Therefore, if the offsets in time are estimated to large, i. e.:

$$\delta t_{up} \; > \; 0, \tag{2.99}$$
$$\delta t_{dn} \; > \; 0, \tag{2.100}$$

the transponder transmits the respective measurement sweep too early. Hence, the signal received by the base station arrives early and the sweeps corresponding to $x_{bs,rx}$ are shifted to the left in figure 2.9. Consequently, f_{up} is increased by a frequency component proportional to δt_{up}, while f_{dn} is decreased by a frequency component proportional to δt_{dn}.

If synchronization errors are taken into account, the frequencies of the low-pass filtered mixed signal in the base station during the measurement upsweep and downsweep are given by:

$$f_{up} = \Delta f_{a,up} - \delta f - (2t_d - \delta t_{up})\,\mu, \tag{2.101}$$
$$f_{dn} = \Delta f_{a,dn} - \delta f + (2t_d - \delta t_{dn})\,\mu. \tag{2.102}$$

Consequently, the distance measurement error depends linearly on δf, δt_{up}, and δt_{dn} [69]. If (2.101) and (2.102) are applied to (2.88) the distance measurement error can be calculated as:

$$\delta d = -\frac{c_{ph}}{4}\left(\delta t_{up} + \delta t_{dn}\right). \tag{2.103}$$

The error depends linearly on the estimation errors of the offsets in time δt_{up} and δt_{dn}. If the offsets in time are estimated too large the transponder transmits the measurement sweeps too early and the distance is measured too short. However, the offset in frequency Δf is used to calculate both offsets in time (2.68), (2.69). The estimation error δf of the offset in frequency, therefore, contributes to the distance measurement error δd as well.

2.4 Relative Motion of the Stations

In the previous sections the distance between the base station and the transponder was assumed to be constant during synchronization and distance measurement. However, if both stations move relatively to each other, the distance changes continuously. Furthermore, the frequency of the received signal is increased by the Doppler frequency shift f_D.

The Doppler frequency shift is given by:

$$f_D \; = \; f_c \left(\sqrt{\frac{c_{ph} + v}{c_{ph} - v}} - 1 \right) \tag{2.104}$$

$$\approx \; f_c \frac{v}{c_{ph}}, \tag{2.105}$$

where f_c is the center frequency of the sweeps and c_{ph} is the phase velocity of the transmitted signals [8, 97]. The relative velocity of the radar stations v is assumed to be positive if the units are moving towards each other. The approximation in (2.105) is valid if the relative velocity of the radar units is small compared to the phase velocity of the radar signals, i. e.:

$$|v| \ll c_{ph}. \tag{2.106}$$

This is true for most applications.

In the following, the influence of the Doppler frequency shift [74, 75] is shown. Furthermore, the effect of the changing distance between both radar stations is analyzed for the first time in the detailed discussion of the synchronization and measurement principle for moving stations below. The novel extensions of the algorithms for synchronization and measurement allow the estimation of the distance of the radar units and their relative velocity.

For the derivations below the relative velocity of the radar units is assumed to be constant during synchronization and distance measurement, i. e.:

$$v = const. \tag{2.107}$$

Furthermore, the distance of the radar units is assumed to be constant during each sweep, but changing from one sweep to another. The distance d_{sync} is defined as the distance of both stations in the middle of the synchronization upsweep. The time-of-flight $t_{d_{sync}}$ is the time-of-flight corresponding to this distance (2.3).

2.4.1 Synchronization of Moving Stations

Figure 2.10 shows the signal constellation in the transponder during synchronization. If both stations move relatively to each other the frequency of the received signal $x_{ts,rx}^D$ is increased by the Doppler frequency shift f_D with respect to the stationary case. Consequently, the frequencies of the low-pass filtered mixed signal f_1^D and f_2^D are decreased by f_D.

The offset in time during the synchronization upsweep is Δt_1. However, the offset in time during the sychronization downsweep Δt_2 differs from Δt_1 if the clock frequencies of both stations do not match (2.55). Furthermore, the distance between the base station and the transponder changes from d_{sync} during the synchronization upsweep to $(d_{sync} - vT_t)$ during the synchronization downsweep. If the stations are moving towards each other the signal received by the transponder $x_{ts,rx}^D$ during the synchronization downsweep arrives earlier with respect to the local signal $x_{ts,lo}$ and the offset in time

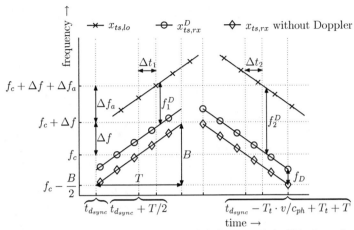

Figure 2.10: Synchronization, Doppler frequency shift (transponder): If both stations move relatively to each other the frequency of the received signal is shifted by the Doppler frequency f_D. The frequency of the received signal without the Doppler frequency shift is also shown. Since the Doppler frequency shift cannot be distinguished from the offset in frequency Δf an error remains after synchronization. Furthermore, the offset in time Δt_1 during the synchronization upsweep differs from the offset in time Δt_2 during the synchronization downsweep due to the relative motion of the stations.

increases by the time-of-flight of the radar signals corresponding to the decrease of the distance.

The scope of (2.55), therefore, is expanded by the additional change in the distance of the stations from the synchronization upsweep to the synchronization downsweep. The offset in time during the synchronization downsweep is then given by:

$$\Delta t_2 = \Delta t_1 \underbrace{- T_t \frac{\Delta f}{f_c}}_{\text{clock}} + \underbrace{\frac{v T_t}{c_{ph}}}_{\text{motion}}, \qquad (2.108)$$

where the changes in the offset in time due to the deviation of the clock frequencies (2.55) and due to the relative motion of the base station and the transponder have been highlighted.

Next, (2.108) is rewritten using the definition of the Doppler frequency shift (2.105). The offset in time during the measurement downsweep is then given by:

$$\Delta t_2 = \Delta t_1 - \frac{\Delta f - f_D}{f_c} T_t. \qquad (2.109)$$

Now the frequencies of the low-pass filtered mixed signal in the transponder during the synchronization upsweep and downsweep are obtained analogously to (2.57) and (2.58). They are described by:

$$f_1^D = (\Delta f - f_D) + \Delta f_a - \mu \Delta t_1, \tag{2.110}$$

$$f_2^D = (\Delta f - f_D) + \Delta f_a + \mu \left(\Delta t_1 - \frac{(\Delta f - f_D)}{f_c} T_t \right). \tag{2.111}$$

Obviously, the Doppler frequency shift f_D cannot be distinguished from the offset in frequency Δf during synchronization.

The frequencies f_1^D and f_2^D are obtained from the spectral analysis if the base station and the transponder move relatively to each other. They are applied to (2.59), (2.68), and (2.69) to calculate estimates Δf^D, Δt_{up}^D, Δt_{dn}^D of the offsets in frequency and in time that need to be corrected to synchronize the measurement sweeps.

However, from (2.59) the offset in frequency is estimated as:

$$\Delta f^D = \frac{f_2^D + f_1^D - 2\Delta f_a}{2 - \frac{T_t}{T} \frac{B}{f_c}} \tag{2.112}$$

$$= \Delta f - f_D. \tag{2.113}$$

Thus, the estimation error δf of the true offset in frequency Δf of the signals of both stations is given by:

$$\delta f = -f_D. \tag{2.114}$$

The offset in time during the measurement upsweep is calculated from (2.68). It is estimated as:

$$\Delta t_{up}^D = \frac{f_2^D - f_1^D}{2\mu} + \frac{T_t}{2} \frac{\Delta f^D}{f_c} - T_p \frac{\Delta f^D}{f_c} \tag{2.115}$$

$$= \underbrace{\Delta t_1 - T_p \frac{\Delta f}{f_c}}_{(2.61)} + T_p \frac{f_D}{f_c} \tag{2.116}$$

$$= \Delta t_{up} + T_p \frac{f_D}{f_c}. \tag{2.117}$$

The true offset in time Δt_{up} is estimated with an estimation error given by:

$$\delta t_{up} = T_p \frac{f_D}{f_c}. \tag{2.118}$$

Similarly, the offset in time for the measurement downsweep obtained from (2.69) is given by:

$$\Delta t_{dn}^D = \frac{f_2^D - f_1^D}{2\mu} - \frac{T_t}{2}\frac{\Delta f^D}{f_c} - T_p\frac{\Delta f^D}{f_c} \tag{2.119}$$

$$= \underbrace{\Delta t_1 - (T_p + T_t)\frac{\Delta f}{f_c}}_{(2.62)} + (T_p + T_t)\frac{f_D}{f_c} \tag{2.120}$$

$$= \Delta t_{dn} + (T_p + T_t)\frac{f_D}{f_c}. \tag{2.121}$$

Thus, the estimation error of the offset in time for the measurement downsweep is described by:

$$\delta t_{dn} = (T_p + T_t)\frac{f_D}{f_c}. \tag{2.122}$$

Since the Doppler frequency shift cannot be distinguished from the offset in frequency of the radar stations during synchronization, the synchronization errors δf, δt_{up}, and δt_{dn} cannot be avoided. They have to be taken into account when the distance and the relative velocity of the units are calculated.

2.4.2 Distance and Velocity Measurement

If the base station and the transponder move relatively to each other, the measurement sweeps cannot be synchronized exactly due to the Doppler frequency shift. However, by evaluating the frequency of the low-pass filtered mixed signal in the base station during the measurement upsweep and downsweep the Doppler frequency shift can be estimated and compensated for. The novel extension of the measurement algorithm presented below then allows to measure the distance and the relative velocity of the radar stations.

Since both stations move relatively to each other, the frequency of the received signal $x_{bs,rx}$ in the base station is once again increased by the Doppler frequency shift. Furthermore, analogously to (2.108), the time-of-flight of the radar signals during the measurement upsweep and the measurement downsweep differs from $t_{d_{sync}}$ during the synchronization upsweep, since the radar stations have moved towards each other with a constant velocity v for a period of T_p and $(T_p + T_t)$, respectively. The time-of-flight of the radar signals during the measurement upsweep and downsweep is given by:

$$t_{d,up} = t_{d_{sync}} - T_p\frac{v}{c_{ph}}, \tag{2.123}$$

$$t_{d,dn} = t_{d_{sync}} - (T_p + T_t)\frac{v}{c_{ph}}. \tag{2.124}$$

Finally, the synchronization errors defined by (2.114), (2.118), and (2.122) have to be taken into account. Figure 2.11 depicts the signal constellation in the base station during the measurement sweeps.

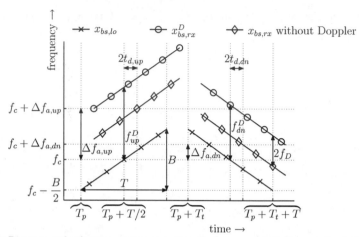

Figure 2.11: Distance and velocity measurement, Doppler frequency shift (base station): The synchronized response of the transponder is shifted by the Doppler frequency shift again. However, by evaluating the frequency of the low-pass filtered mixed signal during the upsweep f_{up}^D and the downsweep f_{dn}^D, the relative velocity of both stations is measured. Then the error due to the Doppler frequency shift can be compensated for and the distance can be calculated as well.

Similarly to (2.101) and (2.102), the frequency of the low-pass filtered mixed signal during the measurement upsweep is given by:

$$f_{up}^D = \Delta f_{a,up} + f_D - \delta f - \mu \left(t_{d_{sync}} + \underbrace{\left(t_{d_{sync}} - T_p \frac{v}{c_{ph}} \right)}_{t_{d,up}} - \delta t_{up} \right) \tag{2.125}$$

$$= \Delta f_{a,up} + 2f_D - 2\mu \left(t_{d_{sync}} - T_p \frac{f_D}{f_c} \right), \tag{2.126}$$

and the frequency of the low-pass filtered mixed signal during the measurement downsweep is described by:

$$f_{dn}^D = \Delta f_{a,dn} + f_D - \delta f + \mu \left(t_{d_{sync}} + \underbrace{\left(t_{d_{sync}} - (T_p + T_t) \frac{v}{c_{ph}} \right)}_{t_{d,dn}} - \delta t_{dn} \right) \tag{2.127}$$

$$= \Delta f_{a,dn} + 2f_D + 2\mu \left(t_{d_{sync}} - (T_p + T_t) \frac{f_D}{f_c} \right). \tag{2.128}$$

Here, f_{up}^D and f_{dn}^D are obtained from the power spectral density of the low-pass filtered mixed signal during the measurement upsweep and downsweep, respectively.

The linear set of equations given by (2.126) and (2.128) is solved with respect to the Doppler frequency shift, i. e.:

$$f_D = \frac{\left(f_{dn}^D - \Delta f_{a,dn}\right) + \left(f_{up}^D - \Delta f_{a,up}\right)}{4 - 2\dfrac{T_t}{T}\dfrac{B}{f_c}}, \qquad (2.129)$$

and the relative velocity of both stations is calculated subsequently from f_{up} and f_{dn} with (2.105). The relative velocity is given by:

$$v = \frac{c_{ph}}{2f_c} \frac{\left(f_{dn}^D - \Delta f_{a,dn}\right) + \left(f_{up}^D - \Delta f_{a,up}\right)}{2 - \dfrac{T_t}{T}\dfrac{B}{f_c}}. \qquad (2.130)$$

The time-of-flight of the radar signals during the synchronization upsweep is also obtained from (2.126) and (2.128) as:

$$t_{d_{sync}} = \frac{T}{4B}\left(\left(f_{dn}^D - \Delta f_{a,dn}\right) - \left(f_{up}^D - \Delta f_{a,up}\right)\right) + \left(T_p + \frac{T_t}{2}\right)\frac{f_D}{f_c}, \qquad (2.131)$$

where the Doppler frequency shift f_D still has to be substituted by (2.129).

From the definition of the time-of-flight (2.3) the distance d_{sync} of both radar stations during the synchronization upsweep can be calculated next. It is described by:

$$d_{sync} = \frac{c_{ph}T}{4B}\left(\left(f_{dn}^D - \Delta f_{a,dn}\right) - \left(f_{up}^D - \Delta f_{a,up}\right)\right) + \left(T_p + \frac{T_t}{2}\right)v. \qquad (2.132)$$

However, if the distance d of both radar units is defined as the average distance of the stations during the measurement sweeps, i. e.:

$$d = \frac{\left(d_{sync} - vT_p\right) + \left(d_{sync} - v\left(T_p + T_t\right)\right)}{2} \qquad (2.133)$$

$$= d_{sync} - v\left(T_p + \frac{T_t}{2}\right), \qquad (2.134)$$

then the distance between the stations is given by:

$$d = \frac{c_{ph}T}{4B}\left(\left(f_{dn}^D - \Delta f_{a,dn}\right) - \left(f_{up}^D - \Delta f_{a,up}\right)\right). \qquad (2.135)$$

This, however, is virtually identical to (2.88) derived for stationary units in section 2.3.2. Therefore, (2.135) is used to estimate the average distance between the base station and the transponder during the measurement sweeps for both the stationary and the non-stationary case. Furthermore, (2.130) provides a novel equation to calculate the relative velocity of the radar stations as well. In the following the superscript-D in f_{up}^D and f_{dn}^D, that has been used to denote the Doppler frequency shift, will be omitted since (2.130) and (2.135) hold for both stationary and moving radar stations.

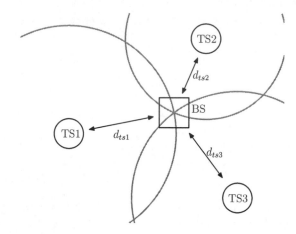

Figure 2.12: Multiple transponders: If the distances to multiple transponders (TS1, TS2, TS3) with known positions are measured, the position of the base station can be obtained by trilateration. The position is given by the intersection of circles centered at the transponders.

2.5 Extension to Multiple Transponder Units

In the previous sections the algorithm to synchronize two radar stations and the algorithm to measure their distance have been discussed. In order to set up a 2D or 3D locating system, the distances and relative velocities of the base station and multiple transponders have to be measured. The concept is shown in figure 2.12.

The transponders (TS1, TS2, TS3) are located at known positions. They serve as reference points for determining the position of the base station. The base station measures its distance to each of the reference transponders. Then, the position of the base station is estimated by trilateration [2, 59].

In a 2D plane the position of the base station is given by the intersection of circles centered at the positions of the transponders. The radii of the circles correspond to the distances d_{ts1}, d_{ts2}, and d_{ts3} between the base station and the transponders. Similarly, the position of the base station is given by the intersections of spheres in a 3D setup.

In the following, three methods to multiplex the distance measurements to multiple transponders are discussed.

2.5.1 Sequential Synchronization and Measurement

A Time Division Multiple Access (TDMA) approach to multiplexing the measurements to multiple transponders can easily be implemented if a communication link between the stations is available.

If a unique station identifier is assigned to each transponder, the base station can address its FSK messages to a single transponder. A measurement request is included in the data sequence transmitted by the base station for pre-synchronization. It contains the unique station identifier of the transponder to be used, e. g. TS1.

The data sequence is received and decoded by all transponders. If a transponder receives a measurement request containing its own station identifier, it switches to synchronization mode. It synchronizes to the synchronization sweeps transmitted by the base station using the algorithm presented in section 2.2.

After synchronization the transponder transmits its measurement sweeps back to the base station. Since the measurement request was addressed to a single transponder, only one synchronized reply arrives at the base station. Consequently, the distance between the base station and the selected transponder, e. g. TS1, and their relative velocity are calculated according to the principle described in section 2.4.2.

The measurement process is repeated until the distances to all transponders are known to the base station. In each measurement cycle a different transponder, e. g. TS2, is selected. If the distances to all transponders have been measured, the position of the base station is estimated by trilateration.

In the setup described above, the maximum distance between each transponder and the base station is given by the maximum measurement range of the 1D distance measurement setup (2.93), (2.92). Since only a single transponder transmits its synchronized reply to the base station in each measurement cycle, there is no interference from other transponders. Furthermore, the number of transponders involved in the measurement process is potentially unlimited.

However, the measurement rate decreases with the number of transponders, since the entire measurement cycle - i. e. pre-synchronization, synchronization, distance and velocity measurement - is repeated for each transponder. Consequently, the approach is not feasible for applications requiring a high measurement rate.

2.5.2 Parallel Synchronization, Sequential Measurement

The measurement rate is increased if all transponders synchronize to the synchronization sweeps of the base station at once. Again the algorithm presented in section 2.2 is used for synchronization.

After synchronization, each transponder transmits its synchronized reply with a dedicated offset in time to the base station. Again, TDMA is used to distinguish the transponders from each other. During the first time slot TS1 transmits its measurement upsweep and downsweep, then the second transponder transmits its measurement sweeps, and so on.

The base station generates local measurement sweeps during all time slots. The local and the received signal are multiplied as before. The low-pass filtered mixed signal is sampled during each time slot and the data are stored temporarily. When the measurement sweeps of all stations have been received the recorded data are evaluated. For each transponder the principle described in section 2.4.2 is used to estimate the distance and relative velocity to the base station. Then the position of the base station is calculated by trilateration.

Again, there is no interference between the transponders and the maximum distance between the transponders and the base station equals the maximum measurement range of the 1D distance measurement setup. Furthermore, the approach can also be used to synchronize multiple transponders to the base station if a communication link is not available. Then sweeps with a low bandwidth have to be used for pre-synchronization as described in section 2.1.5.

The estimation of the offset in frequency during synchronization, however, is only valid for a given time. Therefore, the standard deviation of the measurement results increases as the time elapsing between the synchronization and measurement sweeps increases. If each transponder uses a separate time slot to transmit its reply back to the base station, however, the interval between the synchronization and measurement sweeps increases with the number of transponders. Thus, the number of transponders involved in the measurement must be kept low. Compared to the previous multiplexing concept, additional memory is required in the base station to store the digitized low-pass filtered mixed signal for all transponders. Furthermore, the measurement rate is low since separate measurement sweeps are transmitted and evaluated for each transponder.

2.5.3 Parallel Synchronization and Measurement

In practice, however, a high measurement rate is required for tracking applications. It can be achieved by synchronizing all transponders to the base station simultaneously using the algorithm described in section 2.2. The measurement sweeps of all transponders are then transmitted within the same time slot. An FDMA-like approach is used to separate the signals of the transponders [76].

In section 2.3.3 additional offsets in frequency have been added to the synchronized reply of the transponder to ensure a positive frequency of the low-pass filtered mixed signal in the base station during distance measurement. A large offset in frequency $\Delta f_{a,up}$ is used during the upsweep, while a small offset in frequency $\Delta f_{a,dn}$ is used during the downsweep. Figure 2.13(a) and figure 2.14(a) depict the local signal $x_{bs,lo}$ and the received sweeps $x_{bs,rx}$ during the measurement upsweep and downsweep, respectively. For simplicity, the Doppler frequency shift is not considered here.

The delay of the received signal $x_{bs,rx}$ with respect to the local signal $x_{bs,lo}$ in the base station is given by the round-trip time-of-flight $2t_d$ of the radar signals between both stations. Consequently, the received sweeps shift from left to right in the figures as the distance increases. The frequency of the low-pass filtered mixed signal during the measurement upsweep decreases, while the frequency of the low-pass filtered mixed signal during the measurement downsweep increases as the distance increases. For the received signal one sweep for $d = 0$ and one sweep for $d = d_{max}$ is shown in figure 2.13 and figure 2.14.

The maximum allowable distance d_{max} between the base station and the transponder depends on the additional offsets in frequency $\Delta f_{a,up}$ and $\Delta f_{a,dn}$ of the measurement sweeps (2.93), where $\Delta f_{a,up}$ must not exceed the Nyquist frequency $f_s/2$.

However, the available bandwidth of the low-pass filtered mixed signal in the base station can be used to multiplex the signals of multiple, e. g. K, transponders if the max-

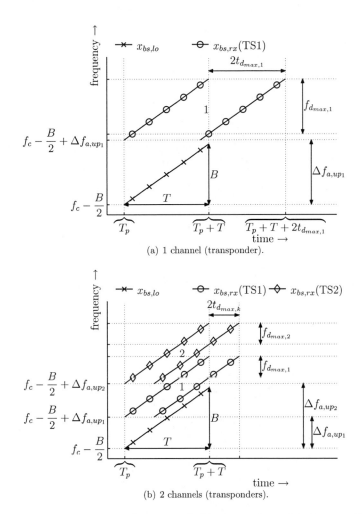

Figure 2.13: iFDMA upsweep (base station): (a) The spectrum of the low-pass filtered mixed signal can be used for a single long measurement channel. (b) Alternatively, it can be divided into non-overlapping channels if the offsets in frequency of the synchronized replies of the transponders, i.e. $\Delta f_{a,up_1}$ and $\Delta f_{a,up_2}$, are chosen appropriately. Then the distances to multiple transponders can be determined with a single measurement sweep.

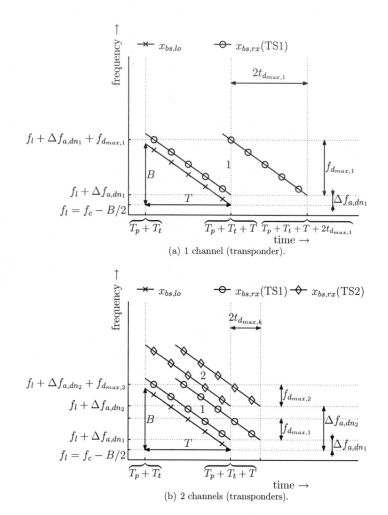

(a) 1 channel (transponder).

(b) 2 channels (transponders).

Figure 2.14: iFDMA downsweep (base station): (a) The spectrum of the low-pass filtered mixed signal can be used for a single long measurement channel. (b) Alternatively, it can be divided into non-overlapping channels if the offsets in frequency of the synchronized replies of the transponders, i.e. $\Delta f_{a,dn_1}$ and $\Delta f_{a,dn_2}$, are chosen appropriately. Then the distances to multiple transponders can be determined with a single measurement sweep.

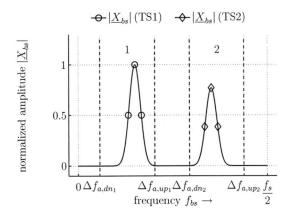

Figure 2.15: iFDMA power spectral density (base station): If the additional offsets in frequency $\Delta f_{a,dn_1}$, $\Delta f_{a,up_1}$, $\Delta f_{a,dn_2}$, and $\Delta f_{a,up_2}$ are chosen appropriately ($\Delta f_{a,dn_2} > \Delta f_{a,up_1}$) the peaks in the power spectral of the low-pass filtered mixed signal in the base station corresponding to the distances to the transponders TS1 and TS2 are well separated from each other. Thus, the distances to both transponders can be measured with the same measurement sweep.

imum allowable distance to each transponder is reduced. An example of the spectrum of the low-pass filtered mixed signal for $K = 2$ transponders is given in figure 2.15.

After synchronization each of the K transponders transmits its synchronized reply within the same upsweep and downsweep, but with different additional offsets in frequency $\Delta f_{a,up_k}$ and $\Delta f_{a,dn_k}$. To divide the spectrum of the low-pass filtered mixed signal in the base station into non-overlapping channels, the additional offsets in frequency for each transponder must be chosen such that:

$$\Delta f_{a,up_k} = \Delta f_{a,dn_k} + f_{d_{max,k}}, \tag{2.136}$$

$$\Delta f_{a,dn_k} > \Delta f_{a,up_{k-1}}, \tag{2.137}$$

where $f_{d_{max,k}}$ denotes the fraction of the frequency of the low-pass filtered mixed signal corresponding to the maximum allowable distance $d_{max,k}$ between the base station and the transponder in channel k. The frequency component $f_{d_{max,k}}$ is given by:

$$f_{d_{max,k}} = 2t_{d_{max,k}}\mu \tag{2.138}$$

$$= \frac{2\mu}{c_{ph}}d_{max,k}. \tag{2.139}$$

The signal constellation in the base station for $K = 2$ transponders is depicted in figure 2.13(b) and figure 2.14(b). Depending on the distance of the radar stations the frequency of the low-pass filtered mixed signal corresponding to transponder TS1 is between $\Delta f_{a,dn_1}$ and $\Delta f_{a,up_1}$. Similarly, the frequency of the low-pass filtered mixed signal corresponding to transponder TS2 is between $\Delta f_{a,dn_2}$ and $\Delta f_{a,up_2}$. Clearly, the peaks in the power spectral density of the low-pass filtered mixed signal in figure 2.15 corresponding to the two transponders are well separated from each other if (2.136) and (2.137) are satisfied. Since the signal components corresponding to the transponders are separated at the intermediate frequency the multiplexing scheme is dubbed Intermediate Frequency Division Multiple Access (iFDMA).

More generally, K received signals are multiplied with the local signal in the base station. The low-pass filtered mixed signal then is a sum of K sinusoidal signals. Similarly to (2.126) and (2.128), the frequency components of the low-pass filtered mixed signal during the measurement upsweep and downsweep are given by:

$$f_{up,k} \;=\; \Delta f_{a,up_k} + 2\frac{v_k}{c_{ph}}f_c - 2\mu\left(t_{d_{sync,k}} - T_p\frac{v_k}{c_{ph}}\right), \tag{2.140}$$

$$f_{dn,k} \;=\; \Delta f_{a,dn_k} + 2\frac{v_k}{c_{ph}}f_c + 2\mu\left(t_{d_{sync,k}} - (T_p + T_t)\frac{v_k}{c_{ph}}\right), \tag{2.141}$$

with $(1 \le k \le K)$.

The low-pass filtered mixed signal is digitized in the base station as before. However, the spectrum of the low-pass filtered mixed signal is evaluated separately for each channel defined by:

$$\Delta f_{a,dn_k} \le f < \Delta f_{a,up_k}, \tag{2.142}$$

and the frequencies $f_{up,k}$ and $f_{dn,k}$ are estimated for each transponder.

Then, the distance d_k from the base station to the k-th transponder and the relative velocity v_k are calculated similarly to (2.135) and (2.130). They are given by:

$$d_k \;=\; \frac{c_{ph}T}{4B}\Big((f_{dn,k} - \Delta f_{a,dn_k}) - (f_{up,k} - \Delta f_{a,up_k}) \Big), \tag{2.143}$$

$$v_k \;=\; \frac{c_{ph}}{2f_c}\frac{(f_{dn,k} - \Delta f_{a,dn_k}) + (f_{up,k} - \Delta f_{a,up_k})}{2 - \dfrac{T_t}{T}\dfrac{B}{f_c}}. \tag{2.144}$$

If the distances to all transponders are known, the position of the base station finally can be estimated by trilateration.

Since the distances and relative velocities to all transponders are measured with a single measurement upsweep and downsweep, the measurement rate is increased significantly compared to the two TDMA concepts presented above.

However, the available bandwidth of the low-pass filtered mixed signal in the base station is split among the transponders. Consequently, the maximum allowable distance between the base station and the transponders is reduced. If the bandwidth of the low-pass filtered mixed signal is split evenly among K transponders, the measurement range reduces by a factor of $1/K$ compared to the range achieved with a single transponder. Furthermore, the system performance can be degraded if the signals of multiple transponders interfere with each other. This is treated in section 4.5.3. Guard bands can be inserted between $\Delta f_{a,up_{k-1}}$ and $\Delta f_{a,dn_k}$ to minimize Adjacent-Channel Interference (ACI) in practice. The measurement range, however, is reduced even further by the guard bands.

The iFDMA scheme presented above has been used for 2D measurement setups [75]. Furthermore, Fuentes Michel et al. used the iFDMA concept to set up a 2D positioning system in which the distances measured to multiple transponders are fused using a simplified version of an extended Kalman filter [24]. Since the distances to all transponders are obtained with a single measurement, a high update rate of the Kalman filter is achieved and the measurement uncertainty of the position of an object is found to be below ±5 cm [24].

2.6 Summary of the Algorithm for Distance and Velocity Measurement

The algorithm for distance and velocity measurement presented in the previous sections can be summarized as follows.

After synchronization the transponder transmits a measurement upsweep and downsweep back to the base station. Both sweeps are received by the base station and multiplied with the corresponding local sweeps. The resulting signal is low-pass filtered and digitized.

The frequency of the low-pass filtered mixed signal depends on the distance d of the base station and the transponder and their relative velocity v. Hence, an FFT algorithm is used to estimate the frequencies of the low-pass filtered mixed signal during the measurement upsweep f_{up} and downsweep f_{dn}.

The frequencies f_{up} and f_{dn} are then used to calculate the relative velocity of the base station and the transponder which is given by:

$$v = \frac{c_{ph}}{2f_c} \frac{(f_{dn} - \Delta f_{a,dn}) + (f_{up} - \Delta f_{a,up})}{2 - \frac{T_t}{T} \frac{B}{f_c}}. \qquad [2.130]$$

The distance d of both stations is defined as the average distance of the units during the measurement sweeps. It is calculated from:

$$d = \frac{c_{ph}T}{4B} \Big((f_{dn} - \Delta f_{a,dn}) - (f_{up} - \Delta f_{a,up}) \Big). \qquad [2.135]$$

The final equations for the velocity (2.130) and distance (2.135) of the stations are repeated here for completeness.

In (2.130) and (2.135) the bandwidth B, the center frequency f_c, the duration T of the measurement sweeps, and the time T_t between the measurement upsweep and downsweep are known system parameters. The phase velocity c_{ph} of the radar signals and the additional offsets in frequency of the measurement upsweep $\Delta f_{a,up}$ and downsweep $\Delta f_{a,dn}$ are known constants as well.

If the relative velocity and distance to multiple transponders are measured, different additional offsets in frequency $\Delta f_{a,up_k}$ and $\Delta f_{a,dn_k}$ are used for each transponder. The frequencies $f_{up,k}$ and $f_{dn,k}$ are estimated for each transponder and the relative velocities v_k and the distances d_k are calculated similarly to (2.130) and (2.135).

2.7 Extended Range of a 1D System

The measurement system presented in this thesis uses (2.130) and (2.135) to calculate the distance and the relative velocity of two radar stations from the frequencies of the low-pass filtered mixed signal in the base station during the measurement upsweep f_{up} and the measurement downsweep f_{dn}. The maximum measurement range of the system depends on the additional offsets in frequency $\Delta f_{a,up}$ and $\Delta f_{a,dn}$ used for the measurement sweeps (2.93). The range is limited by the sampling frequency f_s used to digitize the low-pass filtered mixed signal in the base station (2.92). As long as the distance is below the maximum measurement range f_{up} and f_{dn} are not aliased and the equations derived in the previous sections can be applied.

In principle, however, there is a second mode of operation where the distance of the stations can be measured non-ambiguously. If the distance increases beyond the maximum measurement range, then f_{up} and f_{dn} are aliased. However, if the parameters f_s, $\Delta f_{a,up}$, and $\Delta f_{a,dn}$ are chosen carefully, the distance and the relative velocity of the radar stations are measured with an offset. Thus, the alias of the frequencies f_{up} and f_{dn} can be detected and corrected. Consequently, the measurement range of a 1D system can be increased significantly. This is discussed in detail in the following.

Normal Operation

As the true distance d of both stations increases from $d = 0$, the frequency of the low-pass filtered mixed signal during the measurement upsweep f_{up} decreases (2.126) and the frequency during the measurement downsweep f_{dn} increases (2.128). The measurement system is in normal operation as long as the frequencies f_{up} and f_{dn} are not aliased, i. e.:

$$\Delta f_{a,up} > f_{up} > 0, \tag{2.145}$$

$$\Delta f_{a,dn} < f_{dn} < \frac{f_s}{2}. \tag{2.146}$$

Similarly to (2.126) and (2.128), the following frequencies then are obtained from the spectral analysis:

$$f_{up} = \Delta f_{a,up} + 2\frac{v}{c_{ph}}f_c - 2\frac{B}{T}\left(t_{d_{sync}} - T_p\frac{v}{c_{ph}}\right), \tag{2.147}$$

$$f_{dn} = \Delta f_{a,dn} + 2\frac{v}{c_{ph}}f_c + 2\frac{B}{T}\left(t_{d_{sync}} - (T_p + T_t)\frac{v}{c_{ph}}\right). \tag{2.148}$$

If these frequencies are applied to (2.130) and (2.135) the measured distance d_{meas} and the measured relative velocity v_{meas} of the radar stations match their true distance d and relative velocity v, i. e.:

$$d_{meas} = \underbrace{c_{ph} t_{d_{sync}} - v\left(T_p + \frac{T_t}{2}\right)}_{(2.134)} \tag{2.149}$$

$$= d, \tag{2.150}$$

$$v_{meas} = \frac{c_{ph}}{2f_c} \frac{4\frac{v}{c_{ph}}f_c - 2\frac{B}{T}T_t\frac{v}{c_{ph}}}{2 - \frac{T_t}{T}\frac{B}{f_c}} \tag{2.151}$$

$$= v. \tag{2.152}$$

This is depicted on the left-hand side of the shaded areas in figure 2.16(a) and figure 2.16(b).

If the system is in normal operation, the measured velocity is limited to an interval given by:

$$v_{min} < v_{meas} < v_{max}, \tag{2.153}$$

where v_{min} is the minimum and v_{max} is the maximum possible relative velocity of the radar stations. Both limits are known parameters given by the measurement setup.

Frequency of the Low-Pass Filtered Mixed Signal Aliased

However, as the distance increases f_{up} and f_{dn} will leave the intervals specified by (2.145) and (2.146). If

$$0 > f_{up} > -\frac{f_s}{2}, \tag{2.154}$$

$$\frac{f_s}{2} < f_{dn} < f_s, \tag{2.155}$$

then the aliases f_{up}^A and f_{dn}^A of the frequencies f_{up} and f_{dn} are measured. The aliases are given by:

$$f_{up}^A = -f_{up}, \tag{2.156}$$

$$f_{dn}^A = f_s - f_{dn}. \tag{2.157}$$

Consequently, the frequencies obtained from the spectral analysis change to:

$$f_{up}^A = -\Delta f_{a,up} - 2\frac{v}{c_{ph}}f_c + 2\frac{B}{T}\left(t_{d_{sync}} - T_p\frac{v}{c_{ph}}\right), \tag{2.158}$$

$$f_{dn}^A = f_s - \Delta f_{a,dn} - 2\frac{v}{c_{ph}}f_c - 2\frac{B}{T}\left(t_{d_{sync}} - (T_p + T_t)\frac{v}{c_{ph}}\right). \tag{2.159}$$

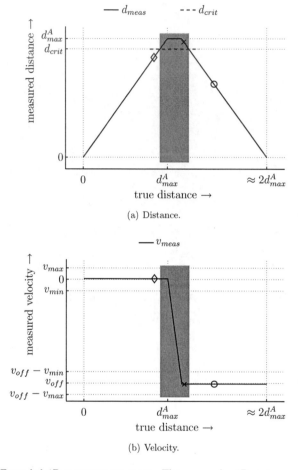

(a) Distance.

(b) Velocity.

Figure 2.16: Extended 1D measurement range: The range of a 1D system can be extended beyond the distance corresponding to $f_s/2$. (a) As the true distance of the stations increases, the measured distance increases as well. If the true distance becomes too large, the frequencies f_{up} and f_{dn} are aliased. Then a false distance value is measured. (b) Aliasing, however, can be detected due to a huge offset in velocity v_{off}.

If these frequencies are applied to (2.130) and (2.135) the distance and relative velocity of the radar stations are measured with an offset, i. e.:

$$d_{meas}^A = -d + \underbrace{\frac{c_{ph}T}{4B}\left(f_s + 2\dot{\Delta}f_{a,up} - 2\Delta f_{a,dn}\right)}_{d_{off}}, \qquad (2.160)$$

$$v_{meas}^A = -v + \underbrace{\frac{c_{ph}}{2f_c}\frac{f_s - 2\Delta f_{a,up} - 2\Delta f_{a,dn}}{2 - \dfrac{T_t}{T}\dfrac{B}{f_c}}}_{v_{off}}. \qquad (2.161)$$

This is shown on the right-hand side of the shaded areas in figure 2.16(a) and figure 2.16(b). The offsets in distance and velocity are given by:

$$d_{off} = \frac{c_{ph}T}{4B}\left(f_s + 2\Delta f_{a,up} - 2\Delta f_{a,dn}\right), \qquad (2.162)$$

$$v_{off} = \frac{c_{ph}}{2f_c}\frac{f_s - 2\Delta f_{a,up} - 2\Delta f_{a,dn}}{2 - \dfrac{T_t}{T}\dfrac{B}{f_c}}. \qquad (2.163)$$

They both depend solely on known system parameters. If f_s, $\Delta f_{a,up}$, and $\Delta f_{a,dn}$ are chosen appropriately, the magnitude of the offset in velocity is much larger than the maximum possible relative velocity of the radar stations.

Hence, if a huge velocity is detected it is concluded that the aliased frequencies have been measured and the true distance and relative velocity of the stations are obtained from:

$$d = d_{off} - d_{meas}^A, \qquad (2.164)$$

$$v = v_{off} - v_{meas}^A. \qquad (2.165)$$

Only one alias is allowed for each of the frequencies f_{up} and f_{dn}. If the distance further increases the frequency of the low-pass filtered mixed signal exceeds the limits given by:

$$f_{up} < -\frac{f_s}{2}, \qquad (2.166)$$

$$f_{dn} > f_s. \qquad (2.167)$$

Then the true distance and relative velocity of the base station and the transponder cannot be measured.

Critical Area

According to (2.163), a significant offset in velocity is only achieved if

$$\frac{f_s}{2} - \Delta f_{a,up} \neq \Delta f_{a,dn}. \qquad (2.168)$$

For the hardware setup presented in the next chapter, for instance, the parameters are chosen such that $f_s/2 - \Delta f_{a,up} = 30\,\text{kHz}$ and $\Delta f_{a,dn} = 50\,\text{kHz}$. The offset in velocity $v_{off} \approx -500\,\text{m/s}$ is then well below the minimum possible relative velocity of the radar stations in typical applications.

Since f_{up} decreases from $\Delta f_{a,up}$ to zero as the distance increases, and f_{dn} increases from $\Delta f_{a,dn}$ to $f_s/2$, both frequencies will be aliased at different distances. The frequency of the low-pass filtered mixed signal in the base station during the measurement downsweep f_{dn} is aliased before the frequency of the low-pass filtered mixed signal during the measurement upsweep f_{up} if:

$$\frac{f_s}{2} - \Delta f_{a,up} < \Delta f_{a,dn}. \tag{2.169}$$

The frequency f_{up} is aliased first if:

$$\frac{f_s}{2} - \Delta f_{a,up} > \Delta f_{a,dn}. \tag{2.170}$$

Either way, there is an area where only one of the frequencies f_{up} or f_{dn} is aliased. In the shaded area in figure 2.16 the frequencies of the low-pass filtered mixed signal obtained from the spectral analysis are either given by f_{up}^A (2.158) and f_{dn} (2.148) or by f_{up} (2.147) and f_{dn}^A (2.159).

If (2.158) and (2.148) are applied to (2.135) the measured distance d_{meas} does not depend on the true distance d, i.e.:

$$d_{meas} = \frac{c_{ph}T}{2B}\left(\Delta f_{a,up} + \frac{v}{c_{ph}}f_c\left(2 - \frac{T_t}{T}\frac{B}{f_c}\right)\right) \tag{2.171}$$

$$= d_{max}^{A_1}. \tag{2.172}$$

Similarly, if (2.147) and (2.159) are used, the measured distance is independent of the true distance d of the stations as well. It is then given by:

$$d_{meas} = \frac{c_{ph}T}{2B}\left(\frac{f_s}{2} - \Delta f_{a,dn} - \frac{v}{c_{ph}}f_c\left(2 - \frac{T_t}{T}\frac{B}{f_c}\right)\right) \tag{2.173}$$

$$= d_{max}^{A_2}. \tag{2.174}$$

Consequently, the distance from the base station to the transponder cannot be measured if only one frequency is aliased. Thus, the shaded area in figure 2.16(a) has to be avoided where

$$d_{meas} = d_{max}^A. \tag{2.175}$$

Therefore, a critical distance d_{crit} is defined. It is chosen such that:

$$d_{crit} < d_{max}^A. \tag{2.176}$$

The system is in normal operation, if a distance below the critical distance is measured and the magnitude of the relative velocity of both stations is small, i.e.:

$$d_{meas} < d_{crit}, \tag{2.177}$$

$$v_{min} < v_{meas} < v_{max}, \tag{2.178}$$

Consequently, (2.150) and (2.152) are used to calculate the distance and the relative velocity of the base station and the transponder.

The frequencies f_{up} and f_{dn} are aliased, if a distance below the critical distance is measured and the magnitude of the relative velocity of both stations is large, i. e.:

$$d_{meas} \quad < \quad d_{crit} \tag{2.179}$$

$$v_{off} - v_{max} < \quad v_{meas} \quad < v_{off} - v_{min}, \tag{2.180}$$

Then, (2.164) and (2.165) are used to calculate the distance and the relative velocity of the base station and the transponder.

However, if a distance above the critical distance is measured, i. e.:

$$d_{meas} > d_{crit}, \tag{2.181}$$

the measurement result is considered invalid. The true distance lies within the shaded areas in figure 2.16. To overcome this problem the base station slightly shifts the frequency of the local measurement sweeps. Thereby, the additional offsets in frequency $\Delta f_{a,up}$ and $\Delta f_{a,dn}$ are modified and the critical area is shifted to other distances. Subsequently, the measurement is repeated.

2.8 Summary

In this chapter the mathematical framework for the distance and velocity measurement concept presented in this thesis has been developed. It has been shown how two radar stations can be synchronized to each other with high precision using FMCW radar signals. After synchronization, their distance can be measured.

The algorithms for synchronization and distance measurement have been analyzed for both stationary and moving radar stations. If the stations move relatively to each other, their relative velocity can be measured as well by evaluating the Doppler frequency shift of the received radar signals. The algorithms which have been derived for synchronization and measurement are summarized in section 2.2 and section 2.6, respectively.

To verify the theory presented in this chapter a measurement system has been designed. It uses the algorithms derived above to synchronize two radar units and measure their distance and relative velocity. The hardware setup of the measurement system is presented in the next chapter. Subsequently, a detailed error analysis of the measurement process is given in chapter 4. Finally, the measurement results are presented in chapter 5.

Chapter 3 — System Design

In chapter 2 a method that utilizes FMCW radar signals for precise distance and velocity measurements has been derived. Based on these algorithms a Local Positioning Radar (LPR) has been developed. In this chapter the design of the novel distance and velocity measurement system is presented.

The chapter starts with an overview of the implementation of the algorithms derived in the previous chapter. Then, key requirements for the signal generator are obtained from the flow chart. Finally, the current hardware configuration is introduced and important system parameters, e. g. the bandwidth and duration of the radar sweeps, are given. These parameters provide the basis for the detailed error analysis in chapter 4 and the measurements presented in chapter 5.

3.1 Overview of the Measurement Cycle

The algorithms presented in chapter 2 are repeated in a cyclic pattern. A complete measurement cycle is comprised of the initialization of the measurement, presynchronization of the base station and the transponder, synchronization of both stations and the actual distance and velocity measurement. An optional time slot for communication concludes the measurement cycle if data has to be exchanged between the stations after the measurement is completed.

Figure 3.1 depicts a flow chart of the measurement cycle. It applies to 1D and 2D measurement setups alike. The distances and relative velocities to up to six transponders are measured by the base station using the iFDMA method presented in section 2.5.3. Multiple base stations measure their distances and relative velocities to the transponders one after another (TDMA). Since the transponders have to synchronize to the base stations sequentially, the measurement rate for each base station is then reduced.

At the beginning of a measurement cycle the measurement is initialized by either the base station or the transponder. During initialization FSK messages are transmitted between the stations. A unique station identifier is used to select a base station and at least one transponder for measurement. The base station then transmits an FSK sequence to the transponder. As discussed in section 2.1.5 the transponder receives the synchronization sequence and adjusts a local timer to this signal. After the synchronization sequence is processed by the transponder the time bases of both radar units coarsely match each other. For the system at hand the offset in time after presynchronization is below 500 ns. The time required for transmitting the FSK sequences is approximately 4.3 ms.

During the next 2 ms the synchronization sweeps are transmitted by the base station. The sweeps are received in the transponder and multiplied with the locally generated

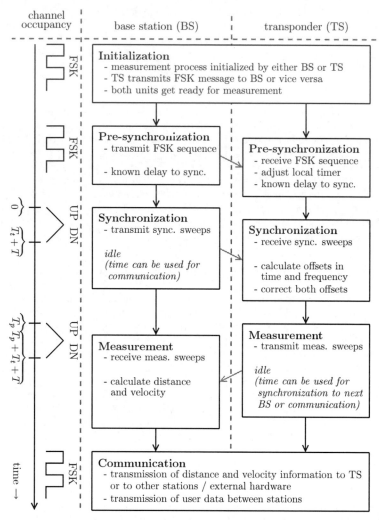

Figure 3.1: Measurement cycle: The measurement process is initialized by either of the stations. The base station then transmits an FSK sequence. The transponder adjusts
a local timer to this signal. After a known delay the synchronization sweeps are
transmitted by the base station. The transponder estimates and corrects its offsets
in time and in frequency before transmitting the measurement sweeps. Then, the
base station calculates the distance and velocity of the stations from the synchronized reply. Finally, data can be exchanged between all stations.

signal. The low-pass filtered mixed signal in the transponder is digitized during the synchronization upsweep and downsweep. The samples are saved temporarily.

The low-pass filtered mixed signal is then evaluated in the transponder. An FFT based algorithm is utilized to estimate the frequency of the signal during the synchronization upsweep and the downsweep. The time required to evaluate the low-pass filtered mixed signal depends on the computational power of the Digital Signal Processor (DSP) which is used for the calculations. For the hardware setup presented here, approximately 6 ms are required to calculate and correct the offsets in time and in frequency of both stations. While the transponder calculates both offsets, the base station is in idle mode. However, the idle time can be used by the base station to communicate with other stations or to transmit synchronization sweeps to another transponder.

After the synchronization sweeps have been transmitted, the base station generates local sweeps for distance and velocity measurement. The delay between the synchronization sweeps and the measurement sweeps is:

$$T_p \approx 8\,\text{ms}. \tag{3.1}$$

The processing time T_p is comprised of the time required by the transponder to estimate and correct the offsets in time and in frequency and the time required for the transmission of the synchronization sweeps.

Since the processing time T_p is a constant system parameter known to both radar stations, the transponder can transmit a synchronized reply back to the base station after the offsets in time and in frequency have been corrected. The transmission of the measurement sweeps again takes approximately 2 ms. In the base station, the received signal is multiplied with the local sweeps, and the low-pass filtered mixed signal is digitized and saved temporarily.

Next, the frequencies of the low-pass filtered mixed signal during the measurement upsweep and downsweep are estimated and the distance and relative velocity of both stations is calculated in the base station. The time required for the calculations depends on the computational power of the DSP and the number of transponders in the system. If only a single transponder is used in a 1D system, it can be as low as 20 ms. While the base station is calculating the distance and relative velocity to each transponder, the transponders are in idle mode. However, the time can be used by each transponder to communicate with other stations or to synchronize to the next base station.

If the measurement results have to be available at all stations they finally are transmitted from the base station to the transponders. Again, an FSK sequence is used for communication. Furthermore, data provided by the user can be transmitted between all LPR units thus eliminating the need for additional communication infrastructure.

If only a single transponder is used with the base station the minimum duration of a complete measurement cycle is:

$$T_c \approx 34.3\,\text{ms}. \tag{3.2}$$

Thus, a measurement rate of up to 29 Hz is obtained. If necessary, the measurement rate can be decreased by adding an additional delay after the communication phase. The measurement rate could be increased significantly if a faster DSP was used for the calculations. Alternatively, the calculation of the FFT of the low-pass filtered mixed

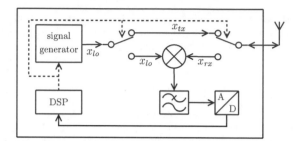

Figure 3.2: Radar module, block diagram: The minimum hardware configuration is comprised
of a DSP for system control and a signal generator to provide the FMCW radar
signals. The output of the signal generator can be transmitted via the antenna or
be fed to a mixer to be multiplied with the received signal. The mixed signal is
then low-pass filtered and digitized.

signal could be implemented in a Field Programmable Gate Array (FPGA) to reduce
the computing time.

3.2 Minimum Hardware Requirements

All LPR modules feature an identical hardware setup. Each station is set to act as a base
station or a transponder in software. The hardware requirements follow directly from
the flow chart depicted in figure 3.1. In this section only hardware components required
for synchronization and distance and velocity measurement are presented. Additional
hardware features are introduced subsequently in section 3.3.

The minimum hardware configuration is depicted in figure 3.2. It includes components for system control, signal generation, signal distribution, and signal processing.

System control: Clearly, a Digital Signal Processor (DSP) is required for the
calculations during synchronization and measurement and for system control. The
processor must allow for a precise control of the timing of the measurement process.
In the transponder the DSP also adjusts the signal generator to compensate for the
calculated offsets in time and in frequency.

Signal generation: The signal generator of each radar module is controlled by the
DSP. The generator provides the FMCW radar signals with linear frequency modulation. The start and stop frequency of the sweeps is adjustable to compensate for any
offset in frequency Δf of the base station and the transponder estimated during synchronization. For FSK communication an additional mode of the generator is required
where the frequency of the output signal is toggled between constant frequencies.

Signal distribution: In transmit (tx) mode the output signal x_{lo} of the signal generator is transmitted directly via the antenna. This is shown in figure 3.2. Alternatively,
the station can be programmed to receive (rx) a signal. The local signal x_{lo} is then fed
to a mixer. There it is multiplied with the received signal x_{rx}. The tx/rx-switches are
controlled by the DSP as well.

Signal processing: In receive mode the output signal of the mixer is low-pass filtered and digitized. The digital values are then sent to the DSP. Here, the recorded data are analyzed. If the station is in transponder mode, the offsets in time and in frequency of both stations are calculated from the frequency of the low-pass filtered mixed signal utilizing the algorithm summarized in section 2.2. The signal generator then is reprogrammed to synchronize the measurement sweeps. If the station is configured as a base station the distance and relative velocity to the transponder are estimated from the power spectral density of the low-pass filtered mixed signal as shown in section 2.6.

3.3 Hardware Setup

Figure 3.3 depicts the printed circuit board of an LPR module. The key components that have been discussed in the previous section, i. e. the DSP [A], the tx/rx-switches [B], the mixer [C], the low-pass filter [D], and the Analog-to-Digital (A/D) converter [E] are labeled in the figure. The DSP and the signal generator are clocked by a single crystal oscillator [F]. Thus, changes in the offset in time of both stations during the processing time due to the deviation of the clock frequencies can be calculated directly from the offset in frequency of the synchronization sweeps (2.60), (2.61), (2.62).

Some additional hardware components are labeled in figure 3.3 as well. An external memory [G] is connected to the DSP. Here the data recorded during the synchronization and measurement sweeps are stored temporarily before they are analyzed. Furthermore, an FSK receiver [H] has been included in the hardware design. It allows for a fast pre-synchronization of the base station and the transponder and the transmission of data, e. g. the measurement results, between all LPR stations.

An antenna switch [I] is used to transmit or receive the radar signals via one of four antenna ports. This allows for measuring 1D distances from the base station to multiple transponders in different directions, even if directional antennas with a narrow 3 dB beam width are used with the system. Furthermore, the orientation of an object can be estimated in 2D applications if the position of the object is estimated using at least two well separated antennas [24, 33].

The transmit power of each station depends on the directional gain of the antennas used with the measurement system. However, the maximum allowable transmit power is given by legal requirements [21, 39, 68]. It must not be exceeded regardless of the antenna gain. Therefore, a step attenuator [J] has been included in the transmit path. It attenuates the level of the transmitted signal by up to 31 dB. Thus, the transmitted signal can be attenuated if antennas with a high directional gain are used. Furthermore, adjacent-channel interference of transponders transmitting their synchronized replies in adjacent measurement channels can be minimized by reducing the transmit power of the transponders close to the base station. This is shown in section 4.5.3.

The signal generator of an LPR station consists of a Phase-Locked Loop (PLL) [K] containing a Voltage Controlled Oscillator (VCO) [L] and a Direct Digital Synthesizer (DDS) [M] used to provide the reference signal for the PLL. The quality of the FMCW signals is most critical for achieving precise distance and velocity measurements. The achievable accuracy of the frequency estimation depends on the bandwidth and the

[M] direct digital synthesizer   [F] crystal oscillator
[L] voltage controlled oscillator   [A] digital signal processor
[K] PLL frequency synthesizer   [G] external memory

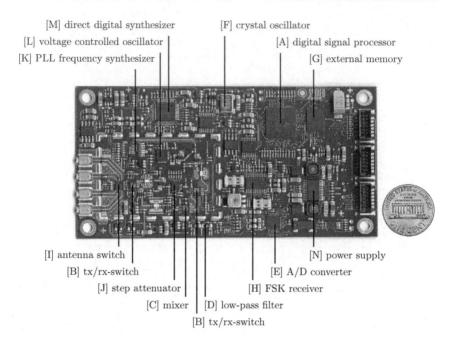

[I] antenna switch   [N] power supply
    [B] tx/rx-switch   [E] A/D converter
        [J] step attenuator   [H] FSK receiver
            [C] mixer   [D] low-pass filter
                    [B] tx/rx-switch

Figure 3.3: Radar module, printed circuit board: The components of the LPR fit on a single sided board, 56 mm x 110 mm. Key hardware components are labeled.

duration of the radar sweeps and on the SNR of the evaluated signal [77,78]. The latter is influenced by the phase noise of the received and the local signal. Furthermore, the parameters of the signal generator determine the precision the offset in frequency is corrected with and the accuracy the sweep rate can be adjusted with. Therefore, special emphasis is placed on the signal generator in the next section.

3.4 Signal Generator

The distance and velocity measurement system presented in this thesis operates in the unlicensed Industrial, Scientific, and Medical (ISM) radio band at 5.8 GHz. Here, a total bandwidth of 150 MHz is available within the frequency range from 5.725 GHz to 5.875 GHz [39, 68]. The signal generator is comprised of two important components, a DDS to generate signals with linear frequency modulation and a frequency of less than 50 MHz, and a PLL with a 5.8 GHz VCO. Since the output of the DDS provides the reference frequency for the PLL, the output frequency of the VCO is proportional to the frequency of the output signal of the DDS.

3.4.1 Direct Digital Synthesizer

Direct digital synthesis is a technique for using digital data processing blocks as a means to generate a frequency- and phase-tunable output signal referenced to a fixed-frequency precision clock source [3]. In essence, the reference clock frequency is "divided down" in a DDS architecture by the scaling factor set forth in a programmable binary tuning word [3]. The tuning word is typically 24-48 bits long which enables a DDS implementation to provide superior output frequency tuning resolution [3].

The frequency tuning word FTW of the DDS used in the current hardware setup is 32 bits long [5]. Therefore, the output frequency of the DDS is given by:

$$f_{dds} = \frac{FTW}{2^{32}} f_{clk},$$ (3.3)

where f_{clk} is the clock frequency of the DDS. The nominal frequency of the crystal oscillator is:

$$f_{clk} = 149.8582 \, \text{MHz},$$ (3.4)

and its frequency stability is ± 25 ppm over all operating conditions.

The DDS also features a linear sweep mode [5]. It is utilized to generate the FMCW signals. Two frequency tuning words, a delta frequency tuning word and a ramp rate word are programmed into the DDS. The frequency tuning words FTW_0 and FTW_1 determine the lower frequency limit and the upper frequency limit of the sweeps, respectively. An additional offset in frequency can be added to the sweeps if FTW_0 and FTW_1 are increased or decreased accordingly.

The sweeps are triggered if the control pin PS0 of the DDS is toggled. If a transition from low to high is detected, an upsweep is triggered and the frequency tuning word is increased from FTW_0 to FTW_1. If a transition from high to low is detected, a downsweep is triggered and the frequency tuning word is decreased from FTW_1 to FTW_0. The sweep rate of the sweeps is given by the delta frequency tuning word $DFTW$ and the ramp rate word RRW. The output frequency of the DDS is increased (upsweep) or decreased (downsweep) automatically by the frequency corresponding to $DFTW$ every $(4RRW)$ clock cycles. The parameters FTW_0, FTW_1, and $DFTW$ are calculated analogously to (3.3).

The DDS is used in two modes. For FSK communication its output frequency is toggled between constant values. During synchronization and measurement the linear sweep mode is utilized to generate sweeps with a center frequency of approximately 44 MHz and a bandwidth of approximately 1 MHz. These signals are used as the reference signal of a PLL.

3.4.2 Phase-Locked Loop

The signal generator of the LPR contains a PLL similar to the standard PLL design described by Banerjee [7]. A block diagram of the signal generator is depicted in figure 3.4. A phase detector compares the output frequency of a 5.8 GHz VCO after it is divided by the N_{pll} divider to a reference frequency. The reference frequency is given by the output frequency of the DDS after it is divided by the R_{pll} divider.

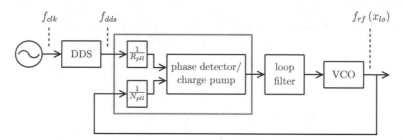

Figure 3.4: Signal generator, block diagram: A DDS is used to generate frequency sweeps
with a center frequency of approximately 44 MHz. The output signal of the DDS
f_{dds} is used as the reference signal of a PLL containing a 5.8 GHz VCO. The
frequency of the output signal of the PLL f_{rf} is proportional to the frequency of
the reference signal f_{dds}. The PLL is programmed to generate frequency sweeps in
the 5.8 GHz ISM band. If the DDS is programmed to toggle its output frequency
f_{dds} between two frequencies the output signal of the PLL f_{rf} can be used for FSK
communication as well.

The average value of the output current of the phase detector is proportional to the
phase error of the compared frequencies [7]. An active loop filter transforms the output
current of the phase detector to the tuning voltage of the VCO. This tuning voltage
adjusts the output phase of the VCO, such that its phase, when divided by N_{pll}, is
equal to the phase of the reference signal [7]. Since the phase is the integral of the
frequency, this implies that the frequencies will also be adjusted. Thus, the output
frequency of the signal generator is proportional to the frequency f_{dds} of the reference
signal provided by the DDS (3.3), i. e.:

$$f_{rf} = \frac{N_{pll}}{R_{pll}} f_{dds}. \tag{3.5}$$

The PLL dividers are set to $N_{pll} = 131$ and $R_{pll} = 1$ to obtain an output signal in the
ISM band at 5.8 GHz from the DDS reference signal with a frequency of approximately
44 MHz. In the following important characteristics of the radar signals are given.

3.5 Characteristics of the Radar Signals

In the ISM radio band at 5.8 GHz a total bandwidth of 150 MHz is available within the
frequency range from 5.725 GHz to 5.875 GHz. The available bandwidth is split into
bandwidth used for FSK communication and bandwidth for the radar sweeps.

3.5.1 FSK Parameters

The number of available FSK channels and hence the bandwidth reserved for FSK
communication depends on the number of LPR stations that have to communicate
with each other in parallel in a particular measurement setup. It can be adjusted in

software. For the first FSK channel the local signal in the receiving station is set to a frequency of:

$$f_{fsk,lo} = 5871\,\text{MHz}. \tag{3.6}$$

The signal of the transmitting station is toggled between the frequencies:

$$f_{fsk,0} = 5860.3\,\text{MHz} - 30\,\text{kHz}, \tag{3.7}$$
$$f_{fsk,1} = 5860.3\,\text{MHz} + 30\,\text{kHz}, \tag{3.8}$$

which correspond to the FSK symbols 0 and 1, respectively. The duration of each symbol is $6\,\mu s$ offering a data rate of approximately 160000 bits per second. In the receiving station the locally generated signal is fed to the mixer, where it is multiplied with the received signal. The mixed signal is band-pass filtered and decoded by an FSK receiver chip. The received data sequence is then processed by the DSP.

In the standard measurement configuration two FSK channels are available. The channels are separated by FDMA. The offset in frequency between adjacent FSK channels is 1 MHz. If only two FSK channels are used a total bandwidth of 132 MHz remains for the synchronization and measurement sweeps. The number of FSK channels, however, can be increased to up to twenty if the bandwidth of the sweeps is reduced. This allows for a flexible system setup for various 1D and 2D applications.

3.5.2 Sweep Parameters

In the standard configuration the bandwidth used for the synchronization and measurement sweeps is:

$$B = 132\,\text{MHz}. \tag{3.9}$$

The center frequency of the sweeps is:

$$f_c = 5793\,\text{MHz}, \tag{3.10}$$

since there is a guard band of 2 MHz to the lower frequency limit of the ISM band. The guard band is required to meet the legal requirements even if the clock frequency differs from its nominal value.

The sweep duration T is determined by the sampling frequency the low-pass filtered mixed signal is digitized with. In the transponder 2048 samples are acquired during each sweep at a frequency of $f_{clk}/72$, in the base station 4096 samples are acquired at a frequency of $f_{clk}/36$, where the clock frequency f_{clk} is given by (3.4). Consequently, the time required for sampling is 0.984 ms in both stations.

The integer delta frequency tuning word is chosen to match the sweep duration and the time required for sampling as precisely as possible. The sweep duration is given by:

$$T = \frac{FTW_1 - FTW_0}{DFTW} \frac{4\,RRW}{f_{clk}}. \tag{3.11}$$

For a bandwidth of 132 MHz the ramp rate word is set to $RRW = 1$. The optimum delta frequency tuning word then is:

$$DFTW = 783, \tag{3.12}$$

and the sweep duration obtained from (3.11) is:

$$T = 0.9845 \, \text{ms}. \tag{3.13}$$

Finally, the sweep rate of the synchronization and measurements sweeps can be calculated from (3.3), (3.5), and (3.11). It is given by:

$$\mu = DFTW \frac{N_{pll}}{2^{32} \, R_{pll} \, (4 \, RRW)} f_{clk}^2, \tag{3.14}$$

and is thus quadratic in f_{clk}, which has been assumed implicitly in (2.35). For the parameters given above the sweep rate is found to be:

$$\mu = 134.083 \, \frac{\text{MHz}}{\text{ms}}. \tag{3.15}$$

3.5.3 Effect of the Distance and Relative Velocity on the Frequency of the Low-Pass Filtered Mixed Signal

The effect of the distance and the relative velocity of the base station and the transponder on the frequency of the low-pass filtered mixed signal can now be quantified. As a rule of thumb the frequency of the low-pass filtered mixed signal (2.83), (2.84) changes by 900 Hz for the standard configuration described above if the distance between both stations changes by 1 m.

However, the Doppler frequency shift (2.105) changes by only 20 Hz if the relative velocity of the stations changes by 1 m/s. For most applications the relative velocity v does not exceed 30 m/s and the Doppler frequency shift f_D does not exceed 600 Hz.

Furthermore, the maximum possible offset in frequency prior to synchronization given by the center frequency of the sweeps and frequency stability of the oscillators is:

$$|\Delta f|_{max} = 2 \cdot 25 \, \text{ppm} \cdot f_c \tag{3.16}$$
$$\approx 290 \, \text{kHz}, \tag{3.17}$$

which is much larger than the maximum Doppler frequency shift.

Consequently, the Doppler frequency shift can be neglected for applications requiring an accuracy of only a few meters. However, for high precision synchronization and distance measurement the Doppler frequency shift has to be taken into account.

3.5.4 Phase Noise of the Radar Signals

The accuracy of the frequency estimation and hence the accuracy of the synchronization and measurement results depends on the SNR of the low-pass filtered mixed signal. It is shown in section 4.5 that the maximum achievable SNR is determined by the phase noise of the FMCW radar signals. Therefore, the phase noise of the output signal of the radar stations is an important system parameter. The phase noise of a 5.8 GHz signal is measured with the spectrum analyzer Rohde & Schwarz FSP [79]. The result is depicted in figure 3.5. The phase noise is between $-80 \, \text{dBc/Hz}$ and $-90 \, \text{dBc/Hz}$ over the entire bandwidth of the active loop filter of the PLL of 200 kHz. It decreases further outside the bandwidth of the loop filter.

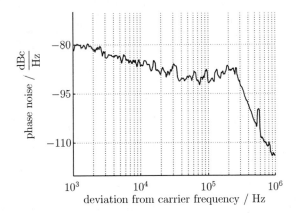

Figure 3.5: Signal generator, phase noise: The output frequency of the system is set to $f_{rf} = 5.8\,\text{GHz}$. The phase noise of the output signal is measured with a spectrum analyzer. It is between $-80\,\text{dBc/Hz}$ and $-90\,\text{dBc/Hz}$ over the entire bandwidth of the loop filter of $200\,\text{kHz}$.

3.5.5 Quantization of the Output Frequency

The output frequency of the signal generator is proportional to the frequency tuning word programmed into the DDS and the clock frequency of the DDS. From (3.3) and (3.5) the frequency tuning word corresponding to a certain output frequency f_{rf} is given by:

$$FTW = 2^{32}\frac{R_{pll}}{N_{pll}}\frac{f_{rf}}{f_{clk}}. \tag{3.18}$$

The output frequency of the signal generator is adjusted by changing the frequency tuning word. Since the frequency tuning word is an integer value, the minimum step size the output frequency of the signal generator can be changed with is:

$$\Delta f_{rf} = \frac{N_{pll}}{R_{pll}}\frac{f_{clk}}{2^{32}}. \tag{3.19}$$

It is obtained by incrementing or decrementing the frequency tuning word by one. For the parameters given above f_{rf} can only be set to multiples of 4.57 Hz. This corresponds to a distance of approximately 5 mm. The quantization error of the output frequency of the signal generator can limit the system performance as shown in section 4.4.3.

3.5.6 Maximum Transmit Power and Measurement Range

The maximum transmit power of each LPR station is 9 dBm. The transmit level is further increased by the gain of the antennas used with the system. However, the transmit level must not exceed the maximum allowable transmit level of 14 dBm effective isotropic radiated power (EIRP) which is given by legal requirements [68]. Hence,

Table 3.1: Important system parameters.

	base station	transponder
clock frequency f_{clk} / MHz	149.8582	
frequency stability / ppm	±25	
center frequency of the sweeps f_c / MHz	5793	
bandwidth of the sweeps B / MHz	132	
duration of the sweeps T / ms	0.9845	
sampling frequency f_s / MHz	4.163	2.081
number of samples per sweep	4096	2048
maximum allowable transmit power / dBm EIRP	14	
measurement range d^+_{max} / km	2.3	
maximum measurement rate / Hz	29	
power consumption / W	4	

the programmable step attenuator is used to attenuate the transmitted signal by up to 31 dB if antennas with a gain of more than 5 dBi are used with the system.

The measurement range of a 1D system strongly depends on the gain of the antennas used with the system. In section 5.1.3, it is shown to be approximately 4 km based on the measured SNR of the received signal if high gain antennas are used with the system. However, the measurement range in normal operation is limited by the frequency the low-pass filtered mixed signal is sampled with in the base station. If the parameters given above are applied to (2.92), the maximum range is found to be:

$$d^+_{max} \approx 2.3 \, \text{km}. \tag{3.20}$$

3.6 Summary

In this chapter the hardware setup of the distance and velocity measurement system LPR has been described. Special emphasis has been placed on the signal generator used to generate the FMCW radar signals. The signal generator consists of a PLL and a DDS to provide the reference signal for the PLL. It can be used for synchronization, distance and velocity measurement and FSK communication alike.

Furthermore, system parameters used for the measurements presented in chapter 5 have been given. They are summarized in table 3.1. Based on these parameters a thorough error analysis of the measurement process is given in the next chapter.

Chapter 4 — System-Theoretic Analysis of the Measurement Process and Identification of Error Sources

In the previous chapters an algorithm to synchronize two radar stations with high precision and to measure their distance and relative velocity has been presented. Furthermore, a hardware implementation of the measurement system has been introduced. The system utilizes FMCW radar signals to synchronize a transponder to a signal transmitted by the base station. The transponder then sends back a synchronized reply and the distance and relative velocity of the radar stations are calculated in the base station.

In the following, a systematic and comprehensive error analysis of the measurement process is given. In the analysis the most important sources of error are identified. It is shown how different parameters affect the accuracy of the measurement results.

The results of the analysis presented in this chapter allow to identify parameters which limit the performance of the current hardware implementation of the measurement system. Furthermore, improvements for future revision of the measurement system can be derived from the results.

The chapter starts with an overview of possible sources of error. The algorithm for frequency estimation then is treated in detail in section 4.2. In subsequent sections the effect of each source of error on the accuracy of the synchronization and measurement results is evaluated in detail. For the analysis theoretical results are complemented by measurement results and simulated data throughout the chapter.

4.1 Classification of Sources of Error

Synchronization and measurement errors can result from the environment the system is working in as well as from the hardware components and the configuration of the measurement system. Figure 4.1 provides an overview of possible sources of error. The most important factors are summarized in the last row of the chart. Other possible sources of error like the floating point precision of the DSP or the quantization error of the Analog-to-Digital Converter (ADC) have not been included in the discussion below since they do not limit the performance of the system presented in chapter 3 in practice.

During synchronization the offsets in time and in frequency of the local signal in the transponder with respect to a signal transmitted by the base station are estimated from the frequency of the low-pass filtered mixed signal in the transponder. Similarly, the distance and relative velocity of the radar stations is calculated from the frequency of the low-pass filtered mixed signal in the base station during measurement. Naturally,

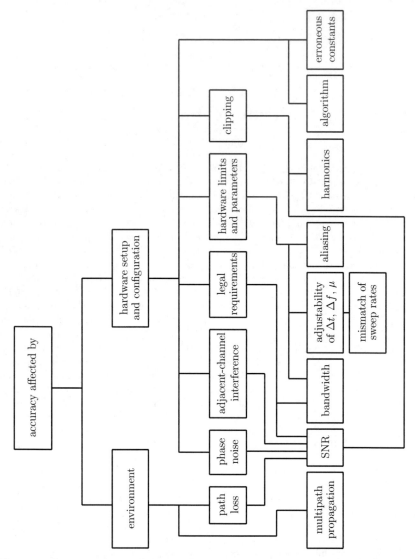

Figure 4.1: Overview of sources of error: The accuracy of the synchronization and measurement results is affected by multipath distortions, the SNR of the low-pass filtered mixed signal, and the bandwidth of the radar sweeps. Furthermore, the precision which the radar stations can be synchronized with is limited by the hardware setup.

the algorithm that is used for the estimation of the frequency of the low-pass filtered mixed signal determines the achievable accuracy of the measurement results.

Key parameters like the sampling frequency f_s and the size of the FFT must be chosen appropriately to achieve optimum performance under the constraints imposed by the hardware setup. For instance, the computational power of the DSP limits the size of the FFT and the frequency f_s the signals can be digitized with. As a consequence, the additional offsets in frequency Δf_a, $\Delta f_{a,up}$, $\Delta f_{a,dn}$ of the synchronization and measurement sweeps must be configured such that aliasing of the frequency of the low-pass filtered mixed signal is avoided. Furthermore, some parameters like the phase velocity of the radar signals are assumed to be known constants for the calculations. In practice, however, these parameters can differ from their nominal values slightly.

If the parameters of the algorithm for frequency estimation are chosen carefully the offsets in time and in frequency of the radar stations can be calculated with high precision. The accuracy with which both offsets can be compensated for, however, is again limited by the hardware setup. The offset in time, for instance, can only be adjusted in multiples of a timer-tick of the DSP. Similarly, the offset in frequency can only be changed by modifying the integer frequency tuning word of the DDS. Furthermore, the sweep rate of the transponder cannot be adjusted to the sweep rate of the base station with the current hardware setup, yielding an additional error if the sweep rates of the radar stations do not match.

It is a well-known fact that the accuracy of the measurement results of an FMCW radar system depends on the bandwidth of the radar sweeps and the SNR of the signals [42, 44]. The bandwidth is limited by legal requirements mainly [21, 39, 68]. Hardware restrictions can apply as well.

On the other hand, the SNR of the low-pass filtered mixed signal depends on a number of factors, most importantly the phase noise of the PLL in the signal generator in each radar module. If iFDMA is used to multiplex the signals of multiple transponders the signals of transponders in adjacent measurement channels can interfere with each other, thus reducing their respective SNR. Furthermore, the SNR is decreased if the distance between the radar stations, and thus the path loss of the radar signals, increases. For short distances, however, clipping of the signal of a transponder close to the base station can cause harmonics in the power spectral densities which are interpreted as signals from transponders in adjacent measurement channels if iFDMA is used.

In the following sections the sources of error identified above are treated in detail. The effect of various parameters on the accuracy of the measurement results is investigated thoroughly. To separate the influences of the environment and the hardware setup line-of-sight only propagation is assumed for the analysis in section 4.2 to section 4.8. However, multipath distortions can degrade the performance of the measurement system significantly. Therefore, they are treated in detail in section 4.9.

4.2 Algorithm for Frequency Estimation

The algorithms for synchronization and distance and velocity measurement presented in chapter 2 rely on the estimation of the frequency of the low-pass filtered mixed signal.

During synchronization the offsets in time and in frequency of the received signal and the local signal are calculated from the frequency of the low-pass filtered mixed signal in the transponder. Similarly, the distance and relative velocity of the radar stations are calculated from the frequency of the low-pass filtered mixed signal in the base station during the measurement sweeps. The estimation of the frequency of the low-pass filtered mixed signal, therefore, is one of the key problems in the calculations.

Various methods for frequency estimation are found in the literature. Kay suggests to use the Fourier transform to estimate the frequency of a single sinusoidal signal embedded in white noise or multiple well separated sinusoidal signals [42]. Especially in dense multipath environments the sinusoidal components of the low-pass filtered mixed signal may not be resolvable by Fourier methods. Subspace based frequency estimation methods, e.g. the Principal Component Autoregressive Method [42, 43] or the MUltiple SIgnal Classification (MUSIC) algorithm [42, 81, 89], can then be used to estimate the frequencies of the sinusoidal signal components. A state space approach for frequency estimation is presented by Rao and Arun [66] and applied to FMCW radar systems by Gulden [31].

However, for the measurement system presented in this thesis the Fourier analysis is used to estimate the frequency of the low-pass filtered mixed signal. It yields excellent measurement results for most measurement scenarios in which the system has been applied so far. In the following the algorithm for frequency estimation is presented in detail. It is comprised of a real valued FFT, a complex valued zoom FFT to improve the resolution of the power spectral density, and a parabolic interpolation of the peak position which has been detected in the power spectral density.

4.2.1 The Fast Fourier Transform

The low-pass filtered mixed signal x_{ts} in the transponder is digitized during the synchronization upsweep and downsweep. During each sweep of duration $T \approx 1\,\mathrm{ms}$, $N_{ts} = 2048$ real valued samples are acquired in the transponder at a sampling frequency of approximately $2\,\mathrm{MHz}$. Similarly, $N_{bs} = 4096$ samples of the low-pass filtered mixed signal x_{bs} are obtained during each measurement sweep of duration $T \approx 1\,\mathrm{ms}$ in the base station. The sampling frequency in the base station is approximately $4\,\mathrm{MHz}$.

After the low-pass filtered mixed signal has been digitized, the FFT, defined as:

$$\underline{X}(m) = \mathcal{FFT}(x(n)) \tag{4.1}$$

$$= \frac{1}{N} \sum_{n=0}^{N-1} x(n) e^{-j2\pi nm/N}, \tag{4.2}$$

is applied to each set of samples to estimate its frequency. Here, x denotes the low-pass filtered mixed signal regardless of whether it has been observed in the base station or the transponder, and N is the number of samples acquired during each sweep. More detailed information on the FFT algorithm is found in the literature [11, 12]. At this point, only the facts relevant to the measurement system LPR are treated in detail.

An exemplary power spectral density measured in an indoor environment is depicted in figure 4.2. The peak magnitude corresponds to the frequency of the low-pass filtered

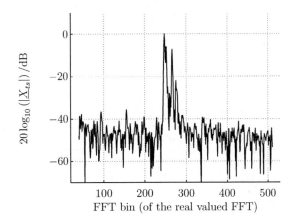

Figure 4.2: Spectral analysis, real valued FFT (measured data, transponder): The low-pass
          filtered mixed signal is sampled for $T \approx 1\,\text{ms}$. Hence, the spacing of the FFT bins
          is $1/T \approx 1\,\text{kHz}$. The resolution of the FFT can be improved by zero padding. To
          minimize the computational effort, a zoom FFT is used to improve the spectral
          resolution in the area around the peak magnitude only.

mixed signal. However, the accuracy of the frequency estimation is severely limited by
the frequency resolution of the FFT which is inversely proportional to the duration the
signal is sampled for [42, 99].

The spacing of the bins in the power spectral density of the low-pass filtered mixed
signal depends on the sampling frequency f_s and the number of samples N acquired
during each sweep. It is given by:

$$f_{bin} = \frac{f_s}{N}. \tag{4.3}$$

Given the sampling parameters specified above, the frequency of the low-pass filtered
mixed signal in either the base station or the transponder can be estimated in multiples
of the spacing of the bins, which is:

$$f_{bin} \approx 1\,\text{kHz}. \tag{4.4}$$

For the sweep parameters given in table 3.1 the spacing of the FFT bins of $1\,\text{kHz}$
corresponds to an estimation of the offset in time in multiples of $3.73\,\text{ns}$ (2.45), an
estimation of the offset in frequency in multiples of $0.5\,\text{kHz}$ (2.46), and finally an estimation of the distance of the stations in multiples of $0.56\,\text{m}$ (2.88). Clearly, only rough
estimates of the frequency of the low-pass filtered mixed signal can be calculated from
the positions of the peak magnitudes in the power spectra during the initial real valued
FFT. The results, therefore, are insufficient for high precision measurements.

However, measurement errors due to the quantization of the frequency estimates can
be reduced if the signal is observed for a longer period of time and more samples are

used when computing the FFT. Alternatively, the number of samples can be increased by zero padding [42,99]. The results of a simulation of the measurement system suggest an adequate frequency resolution if the number of samples is increased by a factor of 4 and the exact position of the peaks in the power spectra is estimated by interpolation of the maxima found by the FFT [70,93]. Both methods will be discussed in the following.

4.2.2 Zero-Padding and Zoom FFT

Figure 4.2 depicts an exemplary power spectrum of the low-pass filtered mixed signal in the transponder during synchronization. The frequency of the low-pass filtered mixed signal is estimated from the position of the peak magnitude in the power spectral density. In the previous section it has been shown that the resolution of the initial real valued FFT is only $1\,\mathrm{kHz}$ which is insufficient for high precision measurements. The resolution of the FFT, therefore, has to be improved to reduce the estimation error due to frequency quantization.

This can be achieved by zero padding. During each synchronization sweep $N_{ts} = 2048$ real valued samples are recorded in the transponder. Similarly, $N_{bs} = 4096$ samples are acquired in the base station during each measurement sweep. If the digitized signal of length N is extended by $3N$ zeros, i. e. if $4N$ values are used in the FFT algorithm, the spacing of the FFT bins is reduced by a factor of 4.

However, the number of operations required to calculate the FFT of a signal of size N is on the order of $N \log_2(N)$ and the time required to calculate the power spectra increases by more than a factor of 4 if $4N$ samples are used instead of N samples [14,95]. More importantly, the size of the FFT is limited to $N = 4096$ on the DSP used for the hardware design presented in section 3 due to memory restrictions. Consequently, zero padding of the digitized values of the low-pass filtered mixed signal to the required length is not a viable option for practical application.

However, it is sufficient to improve the spectral resolution in a small frequency band around the peak position in the spectrum obtained from the initial real valued FFT. This can be achieved by applying a zoom FFT to the results of the real valued FFT [10,25,30,69]. The zoom FFT is comprised of the following steps.

Firstly, the conventional real valued FFT $\underline{X}_{ts}$ (4.2) is applied to the samples recorded during each sweep. The position m_p of the peak magnitude $\max(|\underline{X}_{ts}|)$ is then detected in the power spectral densities. For the exemplary spectrum given in figure 4.2 the position of the peak is $m_p = 247$.

Next, $N = 32$ points centered at m_p are copied from the complex valued results of the initial FFT. The selected section of the spectrum is transformed back to the time domain by applying the Inverse FFT (IFFT) to the copied values. The IFFT is defined as:

$$x(n) = \mathcal{IFFT}(\underline{X}(m)) \tag{4.5}$$

$$= \sum_{m=0}^{N-1} \underline{X}(m) e^{j2\pi mn/N}. \tag{4.6}$$

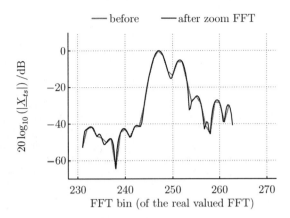

Figure 4.3: Spectral analysis, complex valued zoom FFT (measured data, transponder): 32 points are copied from the complex valued results of the real valued FFT. The center of this region is given by the position of the peak magnitude. The copied values are converted back to the time domain using a complex valued IFFT. Zeros are then added, and the zero padded signal is converted back to the frequency domain using a complex valued FFT. Thus, the resolution of the FFT is improved.

Note, that in (4.6) $\underline{X}(m)$ refers to the $N = 32$ points copied from the original spectrum only. The same notation as in (4.2) has been chosen deliberately to emphasize the similarity of the FFT and the IFFT. In fact, the IFFT can be expressed in terms of the FFT, i.e.:

$$\mathcal{IFFT}\left(\underline{X}(m)\right) = N \cdot \mathcal{FFT}^*\left(\underline{X}^*(m)\right), \tag{4.7}$$

where * denotes the complex conjugate of a complex number. Therefore, a single software implementation of the FFT algorithm [95] can be used to calculate both the FFT and the IFFT.

The signal obtained from the IFFT is zero padded to a length of $N = 128$ and transformed back to the frequency domain using a complex valued FFT. Figure 4.3 depicts the band around the peak before and after the zoom FFT. Before the zoom FFT is applied the spacing of the FFT bins is approximately 1 kHz (gray line). By applying the zoom FFT the resolution is improved by a factor of 4 to approximately 250 Hz (black line).

The spacing of the FFT bins after the zoom FFT is 250 Hz. This corresponds to an estimation of the offset in time in multiples of 0.93 ns (2.45), an estimation of the offset in frequency in multiples of 125 Hz (2.46), and finally an estimation of the distance of the stations in multiples of 0.14 m (2.88).

4.2.3 Interpolation of the Peak Position

The accuracy of the estimation of the frequency of the low-pass filtered mixed signal can be further improved by spectral interpolation. A common method is to fit the three largest adjacent FFT outputs into a parabola [16]. Crinon presents a more sophisticated interpolation scheme as well [16]. However, simulation results indicate that a parabolic fit is sufficient to enhance the accuracy of the frequency estimation to a few Hertz for the measurement system at hand.

For spectral interpolation the position m_z of the peak magnitude $\left| \underline{X}^Z_{mz} \right|$ is detected in the power spectrum obtained from the zoom FFT. A parabola is then fit through the maximum and the FFT outputs $\left| \underline{X}^Z_{mz-1} \right|$ and $\left| \underline{X}^Z_{mz+1} \right|$ obtained at the positions $(m_z - 1)$ and $(m_z + 1)$, respectively. The main lobe in the power spectral density is essentially replaced by a quadratic polynomial. According to Smith this is valid for any practical window transform in a sufficiently small neighborhood about the peak, because the higher order terms in a Taylor series expansion about the peak converge to zero as the peak is approached [86].

The position of the maximum of the parabola, given by:

$$m_{z,ip} = m_z + 0.5 \cdot \frac{\left| \underline{X}^Z_{mz+1} \right| - \left| \underline{X}^Z_{mz-1} \right|}{2 \left| \underline{X}^Z_{mz} \right| - \left| \underline{X}^Z_{mz-1} \right| - \left| \underline{X}^Z_{mz+1} \right|}, \tag{4.8}$$

is then calculated to estimate the interpolated position of the peak in the power spectrum [69,86]. Thereby an accurate estimate of the peak position and hence the frequency of the low-pass filtered mixed signal is obtained.

However, a residual error still remains if the peak does not match the center of an FFT bin [86]. In the following it will be shown how the residual error depends on the size of the zoom FFT, the true position of the peak, and the window function applied to the signal prior to the real valued FFT.

4.2.4 Window Functions

During each sweep the low-pass filtered mixed signal is sampled during a finite time interval only. The interval can be described by:

$$N f_s \approx T, \tag{4.9}$$

where f_s is the sampling frequency, N is the number of samples acquired, and T is the sweep duration. Ideally the Fourier transform of a sinusoidal signal differs only from zero at the frequency of the signal. However, due to the finite observation time interval some signal energy occurs at frequencies different from the frequency of the signal as well. This is commonly referred to as spectral leakage.

The amount of spectral leakage can be reduced by applying window functions to the sampled signal. A detailed analysis of window functions and their application to the FFT is given by Harris [35]. Nuttall presents some additional windows [61]. Figure 4.4 shows some common window functions.

The rectangular window (diamonds) is equal to one during the entire observation interval. It is applied to the sampled data implicitly by observing the signal for a finite

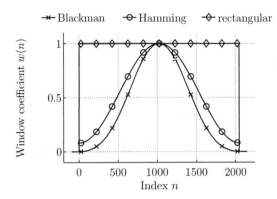

Figure 4.4: Window functions, time domain: Window coefficients of the Blackman window, Hamming window, and the rectangular window.

Table 4.1: Window functions, spectral properties: Highest side lobe level, side lobe falloff, and equivalent noise bandwidth of common window functions [35].

window	highest side lobe level in dB	side lobe falloff in dB/octave	equivalent noise bandwidth F_f in FFT bins
rectangular	-13	-6	1
Hamming	-43	-6	1.36
Blackman	-58	-18	1.73

time. However, the sampled data can be weighted by alternative window functions, e. g. the Hamming window (circles) or the Blackman window (crosses), before the FFT algorithm is applied. The window coefficients of the Hamming window and the Blackman window can be calculated easily [35].

Figure 4.5 depicts the power spectral density of a sinusoidal signal observed for a finite time if a rectangular window (gray line) or a Blackman window (black line) are applied. The peak magnitude is located at the normalized frequency of the signal $n = 500$. The shape of the power spectral density depends on the window function applied to the data. Each window is characterized by the highest level of the side lobes with respect to the peak magnitude, the side lobe falloff, and the equivalent noise bandwidth [35]. Table 4.1 summarizes important properties of common window functions.

The attenuation of the side lobes is important if multiple peaks are present in the power spectral density of the low-pass filtered mixed signal, e. g. if the signals of multiple transponders are multiplexed using the iFDMA scheme presented in section 2.5.3. Furthermore, multiple peaks are obtained in the power spectral densities if the system is used in a multipath environment, which will be treated in detail in section 4.9.1. If multiple peaks are present in the power spectral density the side lobes of each peak in-

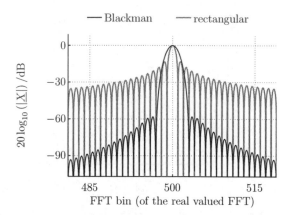

Figure 4.5: Window functions, frequency domain (simulated data): Normalized amplitude spectrum of a sinusoidal signal with normalized frequency 500 if a Blackman window or a rectangular window is applied to the signal.

terfere with the remaining peaks. As the side lobe attenuation increases the distortions of the peak positions due to the interfering side lobes decrease.

However, for the windows considered here the equivalent noise bandwidth F_f increases as well as the side lobe attenuation increases. It is a measure for the width of the main lobe in the power spectral densities [35]. As the equivalent noise bandwidth increases the width of the peaks increases as well. Consequently, the axial resolution of the measurement system diminishes, i. e. the separation in frequency required to resolve two sinusoidal signals increases.

The achievable accuracy of the frequency estimation strongly depends on the window function applied to the data. It is investigated by simulation. The frequency of a single sinusoidal signal is increased from 0 to $f_s/2$. A total of $N = 2048$ samples of the signal are acquired at a sampling frequency of 2 MHz and each of the window functions in figure 4.4 is applied to the digitized values. No additional noise is added to the signals. The FFT algorithm is then used to detect the peak magnitude in the power spectral densities. Finally, the exact position of the peak is estimated by applying the zoom FFT to the band surrounding the peak, and quadratic interpolation of the peak position in the zoomed power spectral density. The estimated peak position is compared to the true position of the peak. Figure 4.6 depicts the simulation results.

If the rectangular window is applied to the sampled data the maximum residual error is approximately $1.2 \cdot 10^{-2}$ FFT bins of the real valued FFT. It is reduced to $0.4 \cdot 10^{-2}$ FFT bins or even $0.07 \cdot 10^{-2}$ FFT bins if the Hamming window or the Blackman window are used, respectively. Hence, the Blackman window is chosen for the implementation of the measurement system. Its highest side lobe level is -58 dB and its equivalent noise bandwidth is 1.73 FFT bins.

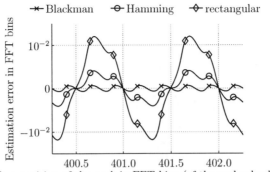

Figure 4.6: Performance of different window functions (simulated data): The performance of the algorithm for frequency estimation depends on the window function applied to the digitized signal. The smallest estimation error is obtained if a Blackman window is used.

Given the sampling parameters specified in table 3.1 the maximum estimation error of the normalized frequency of $0.7 \cdot 10^{-3}$ FFT bins corresponds to an estimation error of the frequency of the low-pass filtered mixed signal of 0.7 Hz (4.4). Since the output frequency of the signal generator of an LPR station can only be adjusted in multiples of 4.57 Hz (3.19), a theoretical estimation error of 0.7 Hz is acceptable for practical application. Given the sweep bandwidth of 132 MHz and the sweep duration of 0.9845 ms this estimation error corresponds to a distance measurement error of 0.8 mm.

4.2.5 Accuracy of the Frequency Estimation

Theoretically, the estimation error of the normalized frequency of the low-pass filtered mixed signal is below $0.7 \cdot 10^{-3}$ FFT bins of the real valued FFT if a Blackman window is applied to the digitized data prior to the FFT, a zoom factor of 4 is used during the zoom FFT, and parabolic interpolation is applied to estimate the peak position. The remaining error could be further reduced if the size of the zoom factor was increased.

Figure 4.7 depicts the residual estimation error if the time domain signal is zero padded to a length of 128 (crosses) and a length of 512 (circles) prior to the complex valued zoom FFT. If the zoom factor is increased from 4 to 16 the maximum estimation error reduces from $0.7 \cdot 10^{-3}$ FFT bins to $0.02 \cdot 10^{-3}$ FFT bins. However, even at a zoom factor of 4 the residual error of the frequency estimation is well below the step size of 4.57 Hz which the output frequency of the signal generator of each LPR unit can be adjusted with. Therefore, a zoom factor of 4 is chosen for the implementation of the measurement system.

Care must be taken near the edges of the power spectral densities. If the frequency f_{ts} of the low-pass filtered mixed signal is close to DC, the side lobes of the peak at

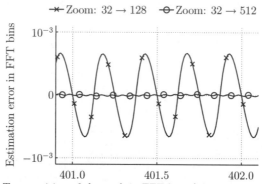

Figure 4.7: Accuracy of the frequency estimation: If a Blackman window is applied to the digitized signal and a zoom factor of 4 is used for the zoom transform (Zoom: $32 \rightarrow 128$) the estimation error after interpolation of the peak position is below $0.7 \cdot 10^{-3}$ FFT bins. If the zoom factor is increased to 16 (Zoom: $32 \rightarrow 512$) the estimation error further decreases.

$(-f_{ts})$ interfere with the peak at f_{ts}. Hence, the maximum estimation error increases. Figure 4.8 depicts the residual estimation error of the peak position for frequencies near DC. As a rule of thumb the interference from the side lobes can be neglected for the sweep and sampling parameters given in table 3.1, if the true position of the peak is larger than 10 FFT bins, i. e. if the frequency of the low-pass filtered mixed signal is larger than approximately 10 kHz.

If the true position of the peak is smaller than 2.5 FFT bins the estimation error of the peak position is between -0.07 FFT bins and 1.04 FFT bins. The axes in figure 4.8, however, have been scaled such that it can be shown that the estimation error reduces below $0.7 \cdot 10^{-3}$ FFT bins quickly.

A similar problem arises if the frequency of the low-pass filtered mixed signal is near $f_s/2$. Aliases of side lobes of the peak corresponding to the frequency of the signal then interfere with the peak. Again, a guard band of approximately 10 kHz should be maintained to the edge of the power spectral density. Similarly, adjacent measurement channels for multiple transponders in the iFDMA setup should be separated by at least 20 kHz to minimize side lobe interference.

4.2.6 Standard Deviation of the Distance and Velocity Measurements

The distance and relative velocity of the radar stations are both calculated from the frequencies of the low-pass filtered mixed signal during the measurement upsweep f_{up} and the measurement downsweep f_{dn}. Therefore, the standard deviation of the velocity

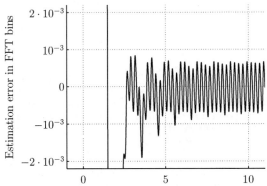

Figure 4.8: Accuracy of the frequency estimation near DC: The algorithm for frequency estimation is less accurate for peak positions near DC. As a rule of thumb the peak position has to be larger than 10 FFT bins to achieve an estimation error of less than $0.7 \cdot 10^{-3}$ FFT bins. If the true position of the peak is smaller than 2.5 FFT bins the estimation error of the peak position is between -0.07 FFT bins and 1.04 FFT bins.

measurements can be estimated from the standard deviation of the distance measurements. This is shown in the following.

Since the same algorithm is used to evaluate the frequency of the low-pass filtered mixed signal during each sweep, it is assumed that the frequencies f_{up} and f_{dn} are estimated with the same standard deviation σ_f. Furthermore, it is assumed that the estimation errors of both frequencies are independent of each other.

From (2.135) the variance of the distance measurements is given by:

$$\text{var}(d) = \left(\frac{c_{ph}T}{4B}\right)^2 2\sigma_f^2, \tag{4.10}$$

where σ_f is the standard deviation of the frequencies f_{up} and f_{dn}. Similarly, the variance of the velocity measurements is obtained from (2.130). It is:

$$\text{var}(v) = \left(\frac{c_{ph}}{2f_c\left(2 - \dfrac{T_t}{T}\dfrac{B}{f_c}\right)}\right)^2 2\sigma_f^2. \tag{4.11}$$

Consequently, the standard deviation of the velocity measurements follows directly from the standard deviation of the distance measurements, i. e.:

$$\text{std}\,(v) \;=\; \frac{2B}{f_c T \left(2 - \dfrac{T_t \, B}{T \, f_c}\right)} \,\text{std}\,(d) \tag{4.12}$$

$$\approx\; \frac{B}{f_c T} \,\text{std}\,(d)\,, \tag{4.13}$$

where $B \ll f_c$ and $T_t \approx T$ are assumed for practical application. For the sweep parameters given in table 3.1 the standard deviation of the velocity measurements is:

$$\text{std}\left(\frac{v}{\text{m/s}}\right) \approx 24\,\text{std}\left(\frac{d}{\text{m}}\right). \tag{4.14}$$

Since the standard deviation of the velocity measurements is proportional to the standard deviation of the distance measurements only the latter will be investigated in detail in the remainder of chapter 4.

4.3 Aliasing and Additional Offsets in Frequency

The frequency of the low-pass filtered mixed signal in the transponder and in the base station is estimated correctly only if it is positive and does not exceed the Nyquist frequency $f_s/2$. Therefore, the additional offsets in frequency of the signal of the transponder during the synchronization and measurement sweeps have to be chosen appropriately to avoid aliasing. In the following a proper choice of the additional offsets is derived.

4.3.1 Additional Offset in Frequency during Synchronization

In section 2.1.3 it has been shown that an additional offset Δf_a can be added to the frequency of the local signal in the transponder to guarantee a positive frequency of the low-pass filtered mixed signal during the synchronization sweeps. The minimum additional offset in frequency is given by:

$$\Delta f_a \geq |\Delta f|_{max} + \mu\,|\Delta t|_{max}\,, \tag{2.42}$$

which is repeated here for convenience. It depends on the sweep rate μ of the radar signals and on the maximum offsets in frequency $|\Delta f|_{max}$ and in time $|\Delta t|_{max}$ of the received signal and the local signal in the transponder prior to synchronization.

The sweep rate μ is a known system parameter. The maximum offset in frequency $|\Delta f|_{max}$ is given by the stability of the oscillators used to clock the radar stations. The maximum offset in time $|\Delta t|_{max}$ on the other hand depends on the precision the radar units have been pre-synchronized with.

The algorithm for pre-synchronization has briefly been described in section 2.1.5. An FSK message is transmitted from the base station to the transponder. In the transponder a local timer is adjusted to the edges in the data sequence. The pre-synchronization

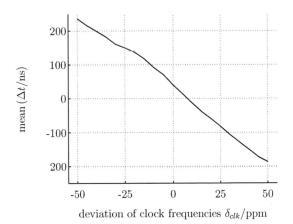

deviation of clock frequencies $\delta_{clk}/$ppm

Figure 4.9: Offset in time after pre-synchronization, mean (measured data, transponder): The mean offset in time Δt after pre-synchronization is within ± 250 ns. It depends on the deviation of the clock frequencies of both stations. The mean offset in time after pre-synchronization is positive if the clock frequency of the transponder $f_{clk,ts}$ is smaller than the clock frequency of the base station $f_{clk,bs}$.

sequence ends with a specific combination of bits. The local synchronization sweeps in the transponder are generated shortly after this combination has been received. They are coarsely synchronized to the synchronization sweeps received from the base station, since the delay between the end of the FSK message and the beginning of the synchronization sweeps is a constant system parameter known to both stations.

The average offset in time between the received signal and the locally generated signal in the transponder after pre-synchronization depends on the deviation δ_{clk} of the clock frequencies of the base station and the transponder. Since the stability of the clock oscillators of the LPR stations is ± 25 ppm, the deviation of the clock frequencies of two stations is within ± 50 ppm. Figure 4.9 depicts the average offsets in time of the stations which have been measured for clock deviations within ± 50 ppm. The measurement results imply a linear dependence of the mean offset in time on the deviation of the clock frequencies.

If the clock frequency of the base station is larger than the clock frequency of the transponder, i. e.:

$$f_{clk,bs} > f_{clk,ts}, \tag{4.15}$$

the deviation of the clock frequencies is negative (2.31).

The algorithm for pre-synchronization is then executed more slowly on the transponder than expected by the base station. The synchronization sweeps received by the transponder, therefore, precede the locally generated sweeps and the offset in time Δt as introduced in section 2.1.1 is positive. Similarly, the mean offset in time is negative if the clock frequency of the transponder is larger than the clock frequency of the

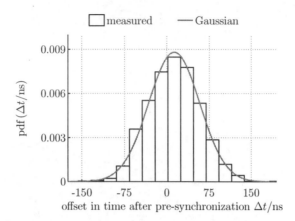

Figure 4.10: Offset in time after pre-synchronization, pdf (measured data, transponder): The probability density function (pdf) of the offset in time Δt after pre-synchronization approximates a Gaussian distribution. The mean depends on the deviation of the clock frequencies of the stations. The standard deviation of the offset in time after pre-synchronzation is less than 50 ns and does not change significantly with the deviation of the clock frequencies.

base station. A small offset in time, however, remains even if the clock frequencies of the stations match since the timer of the transponder can only be adjusted in ticks of approximately 53.4 ns. This will be treated in detail in section 4.4.1.

The mean offset in time after pre-synchronization is well within ±250 ns for all deviations of the clock frequencies of the radar stations. If Δt was assumed to be linear in δ_{clk} and the offset in frequency of both stations was known, e. g. from a previous measurement, the mean value of the offset in time could even be compensated for. However, this has not been implemented in the current software of the LPR.

The probability density function (pdf) of the offset in time after pre-synchronization measured in the transponder is shown in figure 4.10. It approximates a Gaussian distribution. The standard deviation of the offset in time is less than 50 ns and does not depend on the deviation of the clock frequencies of the stations. However, the mean value changes according to figure 4.9.

Now, the additional offset in frequency Δf_a added to the synchronization sweeps of the transponder is calculated from (2.42). It is given by:

$$\Delta f_a \geq 290\,\text{kHz} + 134.083\,\frac{\text{MHz}}{\text{ms}}\,(250\,\text{ns} + 3 \cdot 50\,\text{ns}) \tag{4.16}$$

$$\geq 350\,\text{kHz}, \tag{4.17}$$

where the maximum offset in frequency $|\Delta f|_{max}$ is given by (3.17), the sweep rate μ is defined by (3.15), and the maximum offset in time $|\Delta t|_{max}$ is given by the mean offset in time after pre-synchronization at the maximum deviation of the clock frequencies plus trice the standard deviation.

According to (2.67) the sampling frequency f_s in the transponder has to be larger than $4\Delta f_a$. The sampling frequency is derived from the clock frequency of the transponder by setting various prescalers. In the transponder it is set to:

$$f_s = \frac{f_{clk}}{72} \tag{4.18}$$
$$= 2.081\,\text{MHz}, \tag{4.19}$$

and the additional offset in frequency during synchronization is:

$$\Delta f_a = \frac{f_s}{4} \tag{4.20}$$
$$= 520\,\text{kHz}. \tag{4.21}$$

The frequency of the low-pass filtered mixed signal in the transponder during the synchronization upsweep (2.43) and downsweep (2.44) is then between 170 kHz and 870 kHz. Thus, aliasing is avoided.

4.3.2 Additional Offsets in Frequency during Measurement

During the measurement sweeps aliases of the frequency of the low-pass filtered mixed signal in the base station must be avoided as well. The size of the FFT is limited to $N = 4096$ due to memory restrictions of the DSP. Therefore, the maximum sampling frequency the low-pass filtered mixed signal can be digitized with during each measurement sweep of duration $T \approx 1\,\text{ms}$ is:

$$f_s = \frac{f_{clk}}{36} \tag{4.22}$$
$$= 4.16\,\text{MHz}. \tag{4.23}$$

Again, f_s is derived from the clock frequency of the DSP by setting various prescalers.

As stated in section 2.3.3 the frequency f_{up} of the low-pass filtered mixed signal in the base station during the measurement upsweep (2.83) decreases as the distance to the transponder increases. Similarly, the frequency f_{dn} of the low-pass filtered mixed signal in the base station during the measurement downsweep (2.84) increases as the distance to the transponder increases. Therefore, a large additional offset, given by:

$$\Delta f_{a,up} = \frac{f_s}{2} - 30\,\text{kHz} \tag{4.24}$$
$$= 2.05\,\text{MHz}, \tag{4.25}$$

is added to the frequency of the measurement upsweep transmitted by the transponder and a small additional offset in frequency, given by:

$$\Delta f_{a,dn} = 50\,\text{kHz}, \tag{4.26}$$

is used for the measurement downsweep.

If $\Delta f_{a,up}$ and $\Delta f_{a,dn}$ are chosen according to (4.25) and (4.26), respectively, frequencies of the low-pass filtered mixed signal near DC are avoided. This is required since the noise level in the power spectral density is elevated at those frequencies in practice due to imperfections of electronic components like the mixer. Furthermore, it has been shown in section 4.2.5 that the estimation of the frequency of the low-pass filtered mixed signal is less accurate near the edges of the power spectral density.

The maximum allowable distance between the base station and the transponder can now be calculated from (2.93). It is found to be:

$$d_{max} = 2.23 \, \text{km}. \tag{4.27}$$

As introduced in section 2.7 the frequency of the low-pass filtered mixed signal in the base station will be aliased, if the distance between the stations exceeds the maximum range. From (2.163) the relative velocity is then measured with an offset of:

$$v_{off} \approx -518 \, \frac{\text{m}}{\text{s}}, \tag{4.28}$$

which is beyond the possible relative velocity of the radar units in most applications. Therefore, an implausible magnitude of the relative velocity indicates aliases of the frequencies of the low-pass filtered mixed signal in the base station and, thus, a measurement error.

The range of the measurement system is currently limited by the sampling frequency of the base station. This is one of the limits the hardware setup imposes on the performance of the system. The precision which the stations can be synchronized with is also limited by the electronic components. This is shown in the next section.

4.4 Hardware Limitations during Synchronization

During synchronization the offsets in time Δt and in frequency Δf of the local and the received signal are estimated with high precision in the transponder. Both offsets have to be corrected before the measurement sweeps are transmitted by the transponder. A DSP timer is adjusted to compensate for the offset in time Δt, and the frequency tuning words of the DDS are modified to account for the offset in frequency Δf. Therefore, both offsets can only be adjusted with limited accuracy. This is shown in the following. It is assumed that a constant offset in time Δt has to be corrected to synchronize both measurement sweeps. However, for practical application Δt has to be replaced by Δt_{up} (2.68) and Δt_{dn} (2.69) for the measurement upsweep and downsweep, respectively.

4.4.1 Compensation for the Offset in Time

During synchronization the offset in time is estimated in the transponder with an inaccuracy of a few picoseconds. However, the measurement sweeps of the base station and the transponder are triggered by a timer of the DSP. Hence, the offset in time can be corrected exactly only if it is a multiple of one timer-tick T_{ti}. Otherwise an offset in time Δt_r remains after synchronization.

For the current hardware configuration a 16 bit timer is used to trigger the sweeps. The timer clock is derived from the system clock by a prescaler of eight. At a system clock of 149.8582 MHz a single timer-tick, therefore, is approximately:

$$T_{ti} = 53.4\,\text{ns}. \tag{4.29}$$

The remaining offset in time is in the interval:

$$0 \leq \Delta t_r < 53.4\,\text{ns}. \tag{4.30}$$

It is minimal if the offset in time Δt is a multiple of T_{ti}. The maximum remaining error in time occurs if the remaining offset in time is slightly smaller than a multiple of T_{ti}.

Figure 4.11 depicts continuous measurements of the length of a delay line. During this particular measurement only the measurement downsweep has been used to calculate the length of the delay line. However, the effect of the remaining offset in time Δt_r on the accuracy of the measurement results can be demonstrated just as well.

The true length of the delay line is 24.05 m. It is estimated correctly only if the remaining offset in time is zero, i. e. if the offset in time is a multiple of T_{ti}. This is true for the measurement at $t = 1\,\text{s}$ (circled line). During the next measurement the offset in time is slightly larger and the synchronized reply of the transponder is received with an additional delay corresponding to the remaining offset in time Δt_r. Consequently, a larger distance is measured. The remaining offset in time and hence the distance measurement error increase until Δt_r is larger than or equal to T_{ti} and an additional timer-tick can be corrected.

Thus, the maximum remaining offset in time is one timer-tick, and the round-trip time-of-flight of the radar signals is measured with a maximum error of $T_{ti} \approx 53.4\,\text{ns}$. Hence, the time-of-flight and the corresponding length of the delay line are estimated with a maximum error of $T_{ti}/2 \approx 26.7\,\text{ns}$ and 8 m, respectively.

Clearly, the precision of the DSP timer is insufficient for high precision distance measurements. Next, it is shown that the remaining offset in time, however, can be compensated for by an equivalent offset in frequency. If the remaining offset in time is corrected properly the measurement results are much more accurate. This is shown in figure 4.11 as well (crosses).

4.4.2 Remaining Offset in Time and Equivalent Offset in Frequency

Figure 4.12 shows how the remaining offset in time Δt_r is expressed in terms of an additional offset in frequency Δf_r. The offset in frequency Δf is omitted here for clarity of presentation. Furthermore, the remaining offset in time Δt_r and the time-of-flight t_d are much smaller than the sweep duration T in practice. Therefore, only the beginning of the sweeps is shown in figure 4.12.

During synchronization the frequency sweep of the local signal in the transponder $x_{ts,lo}$ (crosses) is matched to the sweep of the received signal $x_{ts,rx}$ (circles). However, the offset in time Δt can only be adjusted in multiples of a timer-tick T_{ti}. Possible

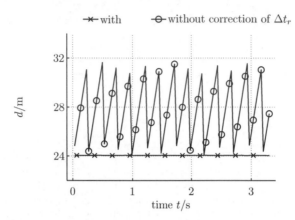

Figure 4.11: Correction of the remaining offset in time (measured data, base station): If the remaining offset in time Δt_r is not compensated for by an equivalent offset in frequency Δf_r, it increases (or decreases, depending on the offset in frequency of both stations) until $|\Delta t_r|$ is larger than or equal to an additional timer-tick T_{ti}. Therefore, the measured distance increases (or decreases) as well until an additional timer-tick can be corrected (circles). If Δf_r is used to compensate for Δt_r the measurement results are much more accurate (crosses).

choices for the sweep of the local signal are marked by diamonds (gray and black). The offset in time Δt is corrected as accurately as possible (diamonds, black) for the signal $x_{ts,tx}$ transmitted during distance measurement.

However, if the offset in time does not match the duration of a timer-tick T_{ti} an offset in time Δt_r remains after the adjustment of the triggers of the sweeps. The remaining offset in time Δt_r is calculated from the offset in time by subtracting T_{ti} until the remaining offset in time is smaller than T_{ti}. It is then converted to an equivalent offset in frequency Δf_r.

From figure 4.12(a) it is apparent that the frequency of the signal $x_{ts,tx}$ transmitted by the transponder has to be increased by an equivalent offset in frequency Δf_r to match the frequency sweeps of the received and the transmitted signal during the measurement upsweep. The equivalent offset in frequency is given by:

$$\Delta f_r = \Delta t_r \frac{B}{T}. \tag{4.31}$$

Similarly, from figure 4.12(b) the frequency of the transmitted signal has to be decreased by Δf_r for the measurement downsweep. For the maximum remaining offset in time of 53.4 ns the equivalent offset in frequency is 7.16 kHz.

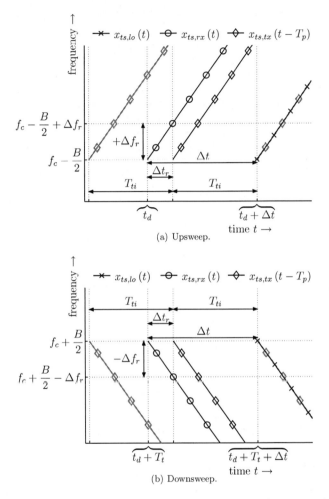

(a) Upsweep.

(b) Downsweep.

Figure 4.12: Equivalence of the offsets in time and in frequency (transponder): The offset in time Δt of the received signal $x_{ts,rx}$ (circles) and the local signal $x_{ts,lo}$ (crosses) is calculated with high precision. However, the timer of the DSP can only be adjusted in timer-ticks $T_{ti} \approx 53.4\,\text{ns}$ (diamonds). Therefore, Δt is corrected as accurately as possible, $x_{ts,tx}$ (black diamonds), and an offset in time Δt_r remains. It is corrected by (a) adding an equivalent offset in frequency Δf_r to the reply of the transponder $x_{ts,tx}$ during the upsweep and (b) subtracting Δf_r during the downsweep. The frequency of the transmitted signal then matches the frequency of the received signal for (a) $t > t_d + \Delta t_r$ and (b) $t > t_d + T_t + \Delta t_r$.

4.4.3 Compensation for the Offset in Frequency

The offsets in frequency Δf and Δf_r are corrected by modifying the frequency tuning words of the DDS. They can be set much more accurately than the offset in time. The accuracy depends on the frequency resolution of the DDS and thus on its bit resolution (3.19).

For the current hardware setup the output frequency of the signal generator can be adjusted in steps of 4.57 Hz. If the frequency tuning word is rounded to the nearest integer value the maximum error in frequency of the synchronized signal due to the resolution of the DDS is 2.29 Hz. Given the sweep parameters in table 3.1 this corresponds to a measurement error of the distance of the radar stations of only 2.5 mm. This is acceptable for practical application.

The effect of the bit resolution of the DDS on the measurements of the length of a delay line is investigated by decreasing the resolution in software. If the 32 bit frequency tuning words are rounded to powers of two, 2^n, the n least significant bits are set to zero and the resolution of the DDS is effectively changed to $(32 - n)$ bit. Consequently, the offset in frequency can only be adjusted in multiples of $(2^n \cdot 4.57\,\text{Hz})$.

The mean of the measured length of the delay line does not change significantly with the bit resolution of the DDS. However, the standard deviation of the distance measurements increases as the DDS resolution is decreased. This is depicted in figure 4.13. The standard deviation changes from 7 mm at a DDS resolution of 32 bit or 31 bit to 266 mm at a resolution of 24 bit. Since the standard deviation of the measurement results remains constant if the resolution of the DDS is changed from 32 bit to 31 bit, the resolution of the DDS does not limit the performance of the measurement system for the current hardware setup as long as the clock frequencies of the radar stations match.

However, if the clock frequencies do not match, the sweep rate of the transponder cannot be adjusted to the sweep rate of the base station due to the resolution of the DDS. As shown in section 4.7.4 the standard deviation of the measurements then increases as the deviation of the clock frequencies, and hence the mismatch of the sweep rates, increases. In the following, it is discussed how the hardware setup can be modified to adjust the sweep rate of the transponder properly.

4.4.4 Adjustment of the Sweep Rate

In section 4.7 it will be shown that the accuracy of the synchronization and measurement results degrades as the mismatch of the sweep rates of the base station and the transponder increases. Since the mismatch of the sweep rates follows directly from the mismatch of the clock frequencies of the radar stations (2.35), the latter must be kept as small as possible to optimize the precision of the system.

In the current hardware setup the frequency stability of the crystal oscillators is ±25 ppm [47]. Hence, the maximum deviation of the clock frequencies of two LPR units is ±50 ppm and the maximum mismatch of the sweep rates is approximately ±100 ppm (2.35). Obviously, the maximum mismatch of the sweep rates can be reduced by replacing the clock oscillators of the stations.

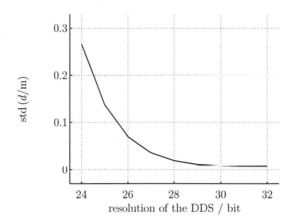

std (d/m)

resolution of the DDS / bit

Figure 4.13: Bit resolution of the DDS (measured data, base station): The length of a delay line is measured and the bit resolution of the DDS is decreased by rounding the tuning words to multiples of 2^n, $n \in [0,8]$, keeping the most significant bits. For the full resolution of the DDS, i.e. 32 bit, and for a resolution of 31 bit the standard deviation of the measured distances is 7 mm. As the resolution further decreases, the standard deviation increases. For a resolution of the DDS of 24 bit the standard deviation is 266 mm. The mean (over 1500 samples) does not change significantly with the resolution of the DDS.

For instance, temperature compensated crystal oscillators (TCXOs) with a frequency stability of ±10 ppm are available [1]. If they are used to clock the LPR units the maximum deviation of the clock frequencies of two stations is ±20 ppm and the maximum mismatch of the sweep rates is reduced to ±40 ppm.

A more precise method to adjust the sweep rate of the transponder, however, is to reprogram the signal generator. According to (3.14) the sweep rate depends on the delta frequency tuning word $DFTW$ programmed into the DDS and the clock frequency f_{clk}. In principle, both parameters could be changed to adjust the sweep rate of the transponder to the sweep rate of the base station. However, with the hardware setup presented in chapter 3 neither of the two parameters can be modified with sufficient accuracy. Hence, a redesign of the clock generator is required if the inaccuracy due to the mismatch of the sweep rates is not acceptable.

The first option is to replace the clock oscillator of the radar stations by a voltage controlled crystal oscillator (VCXO). The clock frequency of the transponder can then be changed directly to compensate for any measured offset in frequency. If the measured offset in frequency is positive the clock frequency of the transponder has to be reduced, if the offset in frequency is negative the clock frequency of the transponder must be increased. After the VCXO has been tuned such that the offset in frequency Δf of the base station and the transponder is close to zero, the clock frequencies of both stations match. Consequently, their sweep rates match as well.

Table 4.2: System parameters for the AD9954 and AD9852.

system parameter	AD9954	AD9852
FTW resolution / bit	32	48
clock frequency / MHz	149.8582	300
center frequency f_c / MHz		5793
sweep bandwidth B / MHz		132
sweep duration T / ms		0.9845
FTW_0	1252954317	41017994697759
FTW_1	1281833306	41963406833276
$DFTW$	783	12803956
$1 / DFTW$	$1.28 \cdot 10^{-3}$	$7.81 \cdot 10^{-8}$
Δf_{rf} / Hz	4.57	$1.4 \cdot 10^{-4}$

The sweep rate of the transponder can also be adjusted by modifying the delta frequency tuning word of the DDS. At the maximum frequency deviation of ± 50 ppm the sweep rates of the base station and the transponder differ by approximately ± 100 ppm. For the current hardware setup the delta frequency tuning word is 783 (3.12). Consequently, the sweep rate can only be adjusted by $1/783 \approx 1.28 \cdot 10^{-3}$ if the delta frequency tuning word is changed by one. Clearly, this is insufficient to correct mismatches of the sweep rate of up to $100 \cdot 10^{-6}$.

However, if the DDS is replaced by a DDS with a resolution of 48 bit [4] the sweep rate of the transponder can be adjusted. The frequency tuning words of the 48 bit DDS are calculated analogously to (3.18), i. e.:

$$FTW = 2^{48} \frac{R_{pll}}{N_{pll}} \frac{f_{rf}}{f_{clk}}, \tag{4.32}$$

where $R_{pll} = 1$ and $N_{pll} = 131$. The delta frequency tuning word obtained from (3.11) with $RRW = 1$ is:

$$DFTW = \frac{FTW_1 - FTW_0}{T} \frac{4RRW}{f_{clk}}. \tag{4.33}$$

In table 4.2 the parameters required to program the 48 bit DDS are compared to the values derived for the current hardware setup in section 3.4 and section 3.5. The delta frequency tuning word of the 48 bit DDS is $DFTW = 1.28 \cdot 10^7$ and the sweep rate of the transponder can then be adjusted with an inaccuracy of less then 0.1 ppm. Since the quantization Δf_{rf} of the output frequency of the signal generator is reduced as well (3.19), replacing the DDS will improve system performance significantly.

However, in the current hardware setup the resolution of the DDS is 32 bit only, and the measurement error due to the possible mismatch of the sweep rates of the base station and the transponder has to be accepted. Furthermore, the accuracy of the measurement results is reduced by the jitter of the clock oscillators, which is discussed next.

4.4.5 Short Term Stability of the Oscillators

The offset in time during the synchronization upsweep (2.60) and the offset in frequency (2.59) of the transponder and the base station are calculated during synchronization. An additional offset in time arises while the data are processed if the clock frequencies of the radar stations do not match (2.61), (2.62). The offset in time arising between synchronization and measurement is calculated from the estimated offset in frequency.

However, the frequency of the oscillators, and hence the offset in frequency of the radar stations, is not constant during the time the data are processed in. The offset in frequency measured during synchronization merely describes the average offset in frequency of the transponder and the base station. The instantaneous offset in frequency drifts continuously and differs slightly from the measured value. Consequently, the period of the clock cycles is not constant and the offset in time arising between synchronization and measurement is estimated with an error. The jitter of the periods of the clock cycles is linked directly to the phase noise of the oscillators [46,55,57].

However, little information is available on the short term stability of the oscillator used in the LPR stations. For a first evaluation of the effect of the short term stability of the oscillator the clock jitter is modeled by a Gaussian random variable. In a simulation of the measurement system a zero mean Gaussian random variable is added to the processing time T_p in the base station. The simulation results are shown in figure 4.14. For a standard deviation of the jitter of more than 20 ps the standard deviation of the distance measurements increases linearly with the standard deviation of the clock jitter. For the oscillators used with the current hardware setup the standard deviation of the jitter arising during the processing time is on the order of 50 ps [46].

The simulation results depicted in figure 4.14, however, are merely a rough qualitative indication of the effect of the clock jitter. Multiple factors contribute to the observed jitter. For instance, the sweeps in the DDS are triggered by toggling a control pin of the DDS. If the threshold at which the transition is detected in the DDS changes from trigger to trigger, an additional jitter is added. However, no data are available on the stability of the trigger threshold.

In general, the standard deviation of the distance measurements increases as the jitter increases. Hence, the jitter must be kept as low as possible to minimize the measurement error. Since the clock jitter is linked to the phase noise of the oscillators, low phase noise oscillators should be used to clock the radar stations. Furthermore, the time between the synchronization and measurement sweeps should be kept as short as possible.

Now that the limitations imposed by the hardware setup have been treated in detail, the effect of some signal parameters, e. g. the SNR of the low-pass filtered mixed signal and the bandwidth of the radar sweeps, will be investigated thoroughly by simulation in the next sections. Each simulation regarding the standard deviation of the distance measurements is run twice. During the first run a jitter of the trigger of the measurement sweeps of approximately 50 ps is assumed. In a second run of the simulation ideal electronic components are assumed and the clock jitter is neglected completely.

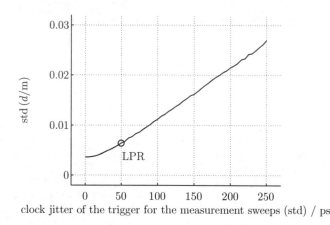

clock jitter of the trigger for the measurement sweeps (std) / ps

Figure 4.14: Clock jitter (simulated data, base station): The standard deviation of the distance measurements increases as the clock jitter increases. To minimize the clock jitter low phase noise oscillators must be used to clock the radar stations. Furthermore, the time between the synchronization and measurement sweeps must be minimized.

4.5 Signal-To-Noise Ratio of the Low-Pass Filtered Mixed Signal

The algorithms for the synchronization of two radar stations and distance measurement presented in this thesis rely on the estimation of the frequency of the low-pass filtered mixed signal in the transponder during the synchronization and in the base station during the measurement sweeps, i.e. on the estimation of the frequency of a sinusoidal signal in noise. The accuracy of the frequency estimation strongly depends on the SNR of the low-pass filtered mixed signal [42], which is given by:

$$\gamma = 20 \log_{10} \left(\frac{A}{\sigma_n} \right). \tag{4.34}$$

Here, A is the amplitude of the sinusoidal signal and σ_n is the standard deviation of the additive noise. Multiple factors contribute to the SNR of the low-pass filtered mixed signal. The most important factors are the path loss during the transmission of the signals, the phase noise of the PLL in the signal generator, interference from transponders transmitting in adjacent measurement channels, and clipping if the signal level is too high. These factors are treated in detail in the following. Subsequently, the dependence of the synchronization and measurement results on the SNR is investigated by simulation.

4.5.1 Path Loss

The amplitude of the low-pass filtered mixed signal x_{ts} in the transponder during synchronization depends on the amplitude of the received signal. The received signal, however, is an attenuated and delayed copy of the signal transmitted by the base station. The attenuation a_{path} is given by the free-space path loss formula, i. e.:

$$a_{path} = 20 \log_{10} \left(\frac{4\pi d}{\lambda} \right). \tag{4.35}$$

It depends on the distance d of the radar stations and the wavelength λ of the radar signals [67, 83], which is defined as:

$$\lambda = \frac{c_{ph}}{f}. \tag{4.36}$$

Consequently, the amplitude of the low-pass filtered mixed signal in the transponder decreases as the distance between the LPR units increases. If the noise level in the receiver is assumed to be constant, the SNR of the low-pass filtered mixed signal decreases as well. At a frequency of $5.8\,\mathrm{GHz}$, i. e. at a wavelength of $5.2\,\mathrm{cm}$, the free-space path loss is approximately $48\,\mathrm{dB}$ at a distance of $1\,\mathrm{m}$. It then increases by $6\,\mathrm{dB}$ each time the distance is doubled.

The measurement sweeps transmitted by the transponder are attenuated on the way to the base station as well. Again, the attenuation is given by the free-space path loss formula (4.35). Since the transponder actively transmits the synchronized reply back to the base station, i. e. a secondary radar is used, the level of the signal received by the base station decreases only with d^2. Hence, the measurement range of the system is increased compared to primary radar systems, where the transponder would simply reflect the signal transmitted by the base station and the level of the received signal would decrease with d^4.

The low-pass filtered mixed signal in the base station can be attenuated further by the anti-aliasing filter between the mixer and the analog-to-digital converter. As the distance between the radar stations increases the frequency of the low-pass filtered mixed signal in the base station during the measurement upsweep decreases (2.83) and the frequency during the measurement downsweep increases (2.84). Hence, the frequency of the low-pass filtered mixed signal during the measurement downsweep is close to the cut-off frequency of the anti-aliasing filter, e. g. $f_s/2$, for large distances. However, for large distances the free-space path loss is also large, and the additional attenuation of up to $3\,\mathrm{dB}$ due to the anti-aliasing filter can limit the maximum range of the measurement system.

4.5.2 Phase Noise of the Phase-Locked Loop

For short distances, i. e. for low attenuations, the SNR of the low-pass filtered mixed signal is not affected by the path loss anymore. This is shown in figure 4.15. The figure depicts power spectral densities measured in the transponder for low attenuations (gray line) and high attenuations (black line) of the received signal.

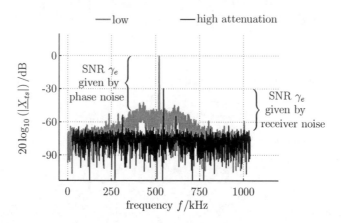

Figure 4.15: Effective SNR (measured data, transponder): For low attenuations of the received signal the effective SNR γ_e of the low-pass filtered mixed signal in the vicinity of the peak is dominated by the phase noise of the PLL in the signal generator (gray line). However, for high attenuations the SNR is given by the receiver noise (black line).

During the spectral analysis only a narrow section of the power spectral density around the peak magnitude is used for the zoom FFT and the subsequent parabolic interpolation. This has been discussed in section 4.2. Clearly, the noise level near the peak is determined by the phase noise of the PLL in the signal generator for low attenuations. Initially, both the peak magnitude in the power spectral density and the noise floor near the peak decrease as the attenuation increases (gray line). Therefore, the SNR observed near the peak magnitude remains constant.

At some point, however, the receiver noise finally exceeds the phase noise of the PLL. If the attenuation increases further the noise floor remains constant and only the peak magnitude reduces. The SNR near the peak magnitude then decreases as the attenuation increases (black line).

In the following, the term effective SNR γ_e is used to characterize the SNR near the peak magnitude in the power spectral density. The effective SNR is determined by the phase noise of the PLL in the signal generator for low attenuations of the received signal, i.e. for short distances between the stations. For high attenuations the effective SNR is limited by the receiver noise and the signal level. The effective SNR is at its maximum as long as the dominant source of noise is the phase noise of the PLL in the signal generator.

The effective SNR can be reduced further if the measurement sweeps of multiple transponders are transmitted to the base station in parallel as introduced in section 2.5.3. This is shown next.

4.5.3 Adjacent-Channel Interference

If the iFDMA approach presented in section 2.5.3 is used to multiplex the synchronized replies of multiple transponders, then their signals are received by the base station at the same time. The sweeps received from the transponders are then multiplied with the local sweep in the base station and the low-pass filtered mixed signal is a sum of sinusoidal signals (2.140), (2.141). However, in practice the phase noise of a strong signal of a transponder close to the base station can reduce the effective SNR of a weak signal of a transponder far away. This is similar to the well-known near-far problem, i. e. the detection of a weak signal in the presence of other stronger signals, encountered in other radar or communication systems [27, 29, 97].

Figure 4.16(a) depicts the power spectral densities of the components of the low-pass filtered mixed signal in the base station during the measurement downsweep if transponder TS4 (gray line) is working in channel $k = 4$ and transponder TS3 is working in channel $k = 3$ (black line). Since the signals of both transponders are received by the base station at the same time, both power spectral densities are superimposed in practice.

The effective SNR γ_e (TS4) of the signal of the nearer transponder TS4 is given by its own phase noise, the peak-to-noise ratio in the power spectral density is approximately 45 dB. However, the effective SNR γ_e (TS3) of the signal of transponder TS3 is also determined by the phase noise of the signal corresponding to transponder TS4, and the peak-to-noise ratio is only 15 dB. Therefore, the effective SNR of transponder TS3 in measurement channel $k = 3$ is impaired severely by interference from transponder TS4 in the adjacent measurement channel $k = 4$.

In figure 4.16(b) the weaker transponder TS6 (black line) is transmitting its synchronized reply in measurement channel $k = 6$. The phase noise of the signal of transponder TS4 (gray line) does not impair the effective SNR γ_e (TS6) of transponder TS6, and the peak-to-noise ratio is 45 dB for both transponders. Measurement channel $k = 5$ is virtually used as a guard band between the measurement channels occupied by the signals of the transponders TS4 and TS6.

In general, adjacent-channel interference can be reduced by sufficiently large guard bands between the measurement channels assigned to the transponders. However, from figure 4.16 a guard band of approximately 750 kHz is required to completely eliminate ACI. Valuable bandwidth of the low-pass filtered mixed signal would be wasted.

Alternatively, ACI is minimized if the transmit power of the transponders is adjusted properly. During synchronization each transponder detects the level of the signal received from the base station. The transponders then use the step attenuator included in the current hardware setup to adjust the level of their synchronized reply: The stronger the level of the signal received by the transponder during synchronization, the higher the attenuation for the synchronized reply. If the signal of transponder TS4 (gray line) in figure 4.16 is attenuated by 30 dB the effective SNR of the signals corresponding to the transponders TS4 and TS6 remains unchanged. The effective SNR of the signal of transponder TS3, however, is increased by 30 dB.

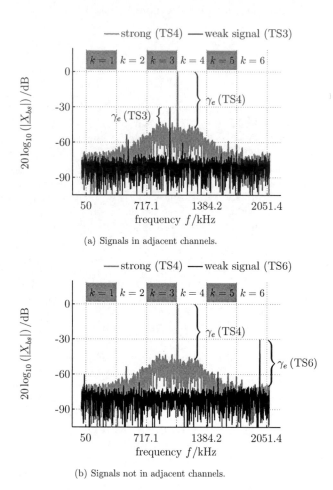

(a) Signals in adjacent channels.

(b) Signals not in adjacent channels.

Figure 4.16: Near-far problem (measured data, base station): If two transponders transmit their measurement sweeps to the base station at the same time, adjacent-channel interference can impair the effective SNR of the signals. (a) The transponder in channel $k = 4$ transmits a strong signal (gray line). The noise of this signal reduces the effective SNR of the weak signal sent by the transponder in the adjacent channel $k = 3$ (black line). (b) Adjacent-channel interference can be mitigated by sufficiently large guard bands or by controlling the transmit power level of each transponder.

If the level of the transmitted signals is not adjusted properly the effective SNR can be reduced further if the level of the received signal is too high and the low-pass filtered mixed signal is clipped.

4.5.4 Clipping of the Low-Pass Filtered Mixed Signal

The low-pass filtered mixed signal is clipped, if its amplitude exceeds the input voltage range of the ADC, which is limited to 1 V peak-to-peak for the current hardware setup. A similar problem arises if the input or output voltage range of an amplifier in the signal path is exceeded. Either way the sampled low-pass filtered mixed signal is not a sinusoidal signal anymore. Consequently, harmonics of the original frequency of the low-pass filtered mixed signal are present in the power spectral density.

Clipping results in two different errors. If the low-pass filtered mixed signal is clipped it can be assumed that the level of the received signal is high and the noise floor near the peak magnitude is determined by the phase noise of the PLL as shown in section 4.5.2. However, if the peak magnitude is clipped and power is shifted to the harmonics of the original frequency of the low-pass filtered mixed signal, the effective SNR clearly is reduced.

Another critical error occurs in the base station if it measures its distance to multiple transponders with a single measurement sweep and the low-pass filtered mixed signal $x_{bs,rx}$ is clipped. Figure 4.17 emphasizes the problem.

In figure 4.17(a) and figure 4.17(b) only a single signal from the transponder in measurement channel $k = 1$ is received. The transponders in the measurement channels $k > 1$ do not transmit their measurement sweeps in this particular setup. In figure 4.17(a) the input voltage range of the ADC and the amplifiers is not exceeded. A single peak in measurement channel $k = 1$ is found in the power spectral density.

However, in figure 4.17(b) the signal level of transponder $k = 1$ is too high. The low-pass filtered mixed signal is clipped and harmonics appear in the measurement channels $k > 1$. The harmonics are detected as well. They are interpreted as signals from the transponders in the measurement channels $k > 1$. If the transponders in those measurement channels were actually transmitting signals the harmonics of the signal component in measurement channel $k = 1$ can interfere with the signal components in the remaining channels. Consequently, erroneous distances to the respective transponders are measured.

Clipping, however, can easily be avoided by adjusting the level of the transmitted signals. The maximum allowable level of the low-pass filtered mixed signal in the receiving station, i. e. the base station, is known from the input voltage range of the ADC. If the level of the low-pass filtered mixed signal is close to this threshold an FSK message is sent to the transmitting station, i. e. the transponder. The transmitting station then reduces its transmit level by increasing the attenuation of the step attenuator to up to 31 dB and the measurement is repeated.

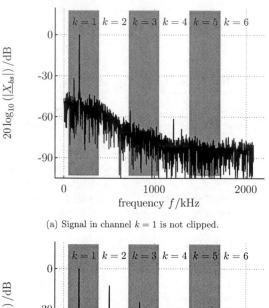

(a) Signal in channel $k = 1$ is not clipped.

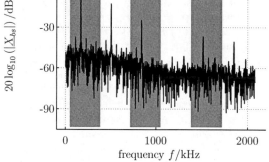

(b) Signal in channel $k = 1$ is clipped.

Figure 4.17: Clipping of the low-pass filtered mixed signal (measured data, base station): Only a single signal from the transponder in measurement channel $k = 1$ is received. (a) If the input voltage range of the ADC and the amplifiers is not exceeded, only a single peak corresponding to the distance to the transponder in measurement channel $k = 1$ is found. (b) If the level of the received signal $x_{bs,rx}$ is too high, the signal is clipped. Then, harmonics of the signal are found in the spectrum as well. They will be detected in the remaining channels as well, although there are no signals from the transponders in the channels $k > 1$.

4.5.5 Accuracy of the Synchronization and Measurement Results

In the previous sections it has been shown that the SNR near the peak magnitude in the power spectral density depends on a number of factors, e.g. the path loss and the phase noise of the PLL in the signal generator. The dependence of the accuracy of the synchronization and measurement results on the SNR is investigated by simulation.

During the simulations Gaussian noise is added to the low-pass filtered mixed signal in the transponder and in the base station. If the standard deviation of the noise is changed, the SNR in the power spectral densities changes as well. The SNR (4.34) is increased from −10 dB to 54 dB. The remaining simulation parameters, e.g. the sweep bandwidth, the sweep duration, and the sampling frequency, match the parameters of the hardware setup presented in chapter 3. For each SNR the simulation is repeated 10000 times. The simulation results are depicted in figure 4.18 and figure 4.19.

The standard deviation of the estimation error of the offset in time during the synchronization upsweep is shown in figure 4.18(a). At an SNR of the low-pass filtered mixed signal of −10 dB it is 530 ps. Theoretically, it reduces to 3.6 ps at an SNR of 54 dB. Similarly, the standard deviation of the estimation error of the offset in frequency, which is shown in figure 4.18(b), decreases from 74 Hz at an SNR of −10 dB to 0.05 Hz at 54 dB. The mean estimation error of the offsets in time and in frequency is zero for all SNRs considered here.

However, the SNR of the low-pass filtered mixed signal cannot be increased indefinitely in practice. In section 4.5.2 it has been shown that the maximum SNR is limited by the phase noise of the PLL in the signal generator. If the SNR according to (4.34) is set to:

$$\max\left(\gamma_e\right) = 20\,\mathrm{dB}, \tag{4.37}$$

spectra with a peak-to-noise ratio similar to the measured power spectral density in figure 4.15 are obtained. The SNR in the simulations then matches the maximum effective SNR which can be achieved with the current hardware setup in practice. The maximum effective SNR has been marked in figure 4.18. At an SNR of 20 dB the offsets in time and in frequency are estimated with a standard deviation of 18 ps and 2.5 Hz, respectively.

The standard deviation of the synchronization errors δt_1 and δf increases as the SNR decreases. Therefore, it can be assumed that the inaccuracy of the distance measurements increases as well. Figure 4.19 depicts the standard deviation of the distance measurements. Again, the SNR of the low-pass filtered mixed signal is increased from −10 dB to 54 dB. The standard deviation of the distance measurements of the LPR (black line) decreases from 100 mm at an SNR of −10 dB to 5.5 mm at an SNR of 54 dB. If the clock jitter due to the short term stability of the oscillators is neglected (gray line) the standard deviation reduces by another 4 mm for high SNRs.

In figure 4.19 the maximum SNR of the low-pass filtered mixed signal, which can be achieved with the current hardware setup in practice, has been marked again. At the maximum effective SNR of 20 dB the standard deviation of the distance measurements is 6.3 mm. The mean estimation error of the distance is zero for all SNRs considered here.

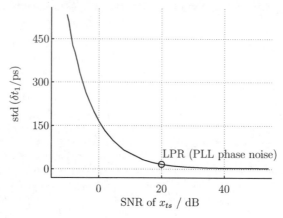

(a) Estimation error δt_1 of the offset in time.

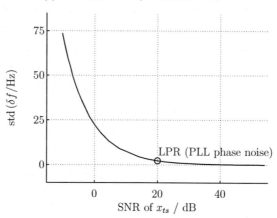

(b) Estimation error δf of the offset in frequency.

Figure 4.18: Dependence of the synchronization results on the SNR (simulated data, transponder): The standard deviation of the estimation error of the offsets in time (a) and in frequency (b) decreases as the SNR of the low-pass filtered mixed signal increases. At an SNR of -10 dB both offsets are estimated with a standard deviation of 530 ps and 74 Hz, respectively. Theoretically, the standard deviations decrease to 3.6 ps and 0.05 Hz if the SNR is increased to 54 dB. However, in practice the maximum SNR of the low-pass filtered mixed signal is determined by the phase noise of the PLL. It is 20 dB. Consequently the offsets in time and in frequency are estimated with a minimum standard deviation of 18 ps and 2.5 Hz, respectively.

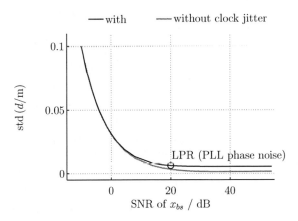

Figure 4.19: Dependence of the distance measurement results on the SNR (simulated data, base station): The standard deviation of the distance measurements decreases as the SNR of the low-pass filtered mixed signal increases. At an SNR of $-10\,$dB the standard deviation is $100\,$mm. Theoretically, it decreases to $5.5\,$mm if the SNR is increased to $54\,$dB (black line). If the clock jitter is neglected it reduces by another $4\,$mm for high SNRs (gray line). However, in practice the maximum SNR of $20\,$dB is given by the phase noise of the PLL, and the standard deviation of the distance measurements is $6.3\,$mm.

The accuracy of the synchronization and measurement results, however, is not affected by the SNR only. It depends on the bandwidth of the radar sweeps as well. This is shown in the next section.

4.6 Bandwidth of the Radar Signals

It is a well-known fact that the accuracy of the measurement results of an FMCW radar system depends on the bandwidth of the radar sweeps [44]. The effect of the bandwidth of the FMCW radar signals on the accuracy of the synchronization and measurement results of the measurement system at hand is investigated by simulation as well.

For the simulations an SNR of the low-pass filtered mixed signal of $20\,$dB is assumed. This corresponds to the maximum achievable SNR for the current hardware setup. The bandwidth of the sweeps is increased from $20\,$MHz to $500\,$MHz. The remaining simulation parameters, e.g. the sweep duration and the sampling frequency, match the parameters of the hardware setup presented in chapter 3. For each bandwidth the simulation is repeated 10000 times. The synchronization results are depicted in figure 4.20.

The standard deviation of the estimation error δt_1 of the offset in time is shown in figure 4.20(a). According to (2.60) it is proportional to $1/B$ if a constant estimation error of the frequencies f_1 and f_2 is assumed. Consequently, the standard deviation of the estimation error δt_1 of the offset in time during the synchronization upsweep reduces from 122 ps at a bandwidth of 20 MHz to merely 5 ps at a bandwidth of 500 MHz.

Figure 4.20(b) depicts the standard deviation of the estimation error δf of the offset in frequency. The standard deviation of 2.5 Hz is nearly independent of the sweep bandwidth. This follows directly from (2.59), where the denominator changes by less than 5 % for $T \approx T_t$, 20 MHz $\leq B \leq$ 500 MHz, and $f_c \approx 5.8$ GHz.

In practice the bandwidth of the FMCW radar signals is limited by legal requirements. The measurement system at hand is working in the ISM band at 5.8 GHz. There, a total bandwidth of 150 MHz is available. A bandwidth of 18 MHz is required for FSK communication and the guard bands near the edges of the ISM band. Therefore, a bandwidth of only 132 MHz remains for the radar sweeps. This has been marked in figure 4.20. At a bandwidth of 132 MHz the offset in time Δt_1 and the offset in frequency Δf are estimated with a standard deviation of 18 ps and 2.5 Hz, respectively.

The standard deviation of the estimation error of the offset in time reduces as the bandwidth of the FMCW radar signals is increased. Therefore, it can be assumed that the inaccuracy of the distance measurement results reduces as well. Figure 4.21 depicts the standard deviation of the distance measurements. Again, the bandwidth of the radar sweeps is changed from 20 MHz to 500 MHz in the simulation.

If the jitter of the sweep triggers is neglected (gray line) the standard deviation of the distance measurements reduces from 23.7 mm at a bandwidth of 20 MHz to 1.1 mm at a bandwidth of 500 MHz. It is then nearly proportional to $1/B$. However, if the bandwidth exceeds 250 MHz the accuracy is limited by the SNR of the low-pass filtered mixed signal.

In practice, however, the jitter of the sweep triggers must be considered as well. If the clock jitter is included in the simulations (black line) the standard deviation of the distance measurements decreases from 23.7 mm at a sweep bandwidth of 20 MHz to 5.4 mm at a bandwidth of 500 MHz. For the current system parameters the standard deviation is approximately 6.3 mm at a bandwidth of 132 MHz.

Clearly, the standard deviation of the distance measurements can only be reduced significantly by increasing the bandwidth of the measurement sweeps if the clock jitter is reduced as well. Consequently, better oscillators have to be used in practice to clock the radar stations.

Increasing the bandwidth of the FMCW radar signals, however, will improve the resolution of multipath components contained in the received signal. Thus, measurement errors due to multipath distortions can be reduced. In section 4.9.2 it will be shown that this is the main benefit of increasing the bandwidth of the sweeps.

In the error analysis above it has been assumed that the sweep rate of the transponder is adjusted to the sweep rate of the base station. The sweep rates of both stations, however, cannot be matched with the current hardware setup. Therefore, the effect of a possible mismatch of the sweep rates is analyzed in detail in the next section.

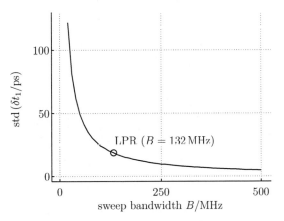

(a) Estimation error δt_1 of the offset in time.

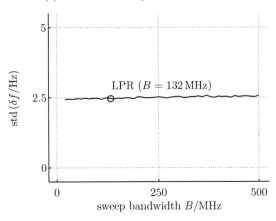

(b) Estimation error δf of the offset in frequency.

Figure 4.20: Dependence of the synchronization results on the bandwidth of the sweeps (simulated data, transponder): (a) The standard deviation of the estimation error δt_1 of the offset in time during the synchronization upsweep decreases from 122 ps to 5 ps as the sweep bandwidth is increased from 20 MHz to 500 MHz. For the current configuration of the LPR a standard deviation of 18 ps is achieved at a bandwidth of 132 MHz. (b) However, the standard deviation of the estimation error δf of the offset in frequency is nearly independent of the sweep bandwidth. It is approximately 2.5 Hz.

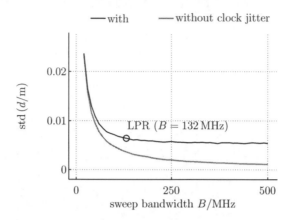

Figure 4.21: Dependence of the distance measurement results on the bandwidth of the sweeps (simulated data, base station): The standard deviation of the distance measurements decreases as the bandwidth of the sweeps increases. At a bandwidth of 20 MHz the standard deviation is 23.7 mm. It reduces to 1.1 mm at a bandwidth of 500 MHz if the clock jitter is neglected (gray line). In practice, the clock jitter must be considered as well (black line). The standard deviation of the distance measurements is then 5.4 mm at a bandwidth of 500 MHz and 6.3 mm at 132 MHz.

4.7 Mismatch of the Sweep Rates of the Radar Stations

In section 2.1.2 and section 2.3.1 a general mathematical description of the frequency of the low-pass filtered mixed signal in the transponder during the synchronization sweeps and in the base station during the measurement sweeps has been given. It has then been assumed that the sweep rates of both radar stations match to simplify the equations for practical application.

However, in section 4.4.4 it has been shown, that the sweep rate of the transponder cannot be adjusted to the sweep rate of the base station with the current hardware setup. If the signal generator of the current hardware setup is modified, e. g. if a VCXO is used to clock the stations or if the resolution of the DDS is increased, the sweep rates could be matched. However, for the hardware setup presented in chapter 3 a possible mismatch of the sweep rates of both stations must be included in the error analysis.

In the following, the mismatch of the sweep rates is analyzed analytically. It is shown subsequently how the deviation of the sweep rates of both stations affects the power spectral densities. Finally, the accuracy of the synchronization and measurement results with respect to the mismatch of the sweep rates is investigated by simulation.

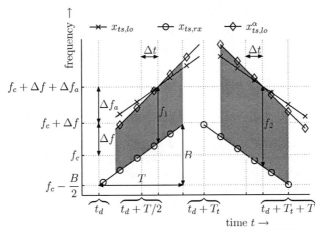

Figure 4.22: Synchronization, different sweep rates (transponder): If the sweep rates of the received and the locally generated signal do not match, the frequency of the low-pass filtered mixed signal in the transponder is not constant during the synchronization upsweep or downsweep (shaded areas). This can be modeled as a chirp around the frequencies f_1 and f_2. Here, the sweep rate of the transponder is larger than the sweep rate of the base station.

4.7.1 Linear Chirps of the Low-Pass Filtered Mixed Signal

The mismatch of the sweep rates of the radar stations severely affects the frequency of the low-pass filtered mixed signal in the transponder during the synchronization sweeps. The frequency of the low-pass filtered mixed signal in the base station during the measurement sweeps is affected in a similar way.

Synchronization in the Transponder

Figure 4.22 depicts the frequency of the local signal in the transponder during synchronization and the signal received from the base station. If the sweep rate of the local sweep in the transponder $x_{ts,lo}$ matches the sweep rate of the received signal $x_{ts,rx}$ the frequency of the low-pass filtered mixed signal is constant during the synchronization upsweep and synchronization downsweep.

However, the frequency of the low-pass filtered mixed signal changes linearly during each synchronization sweep if the sweep rate of the local signal in the transponder $x_{ts,lo}^{\alpha}$ does not match the sweep rate of the signal received from the base station. The respective areas for the synchronization upsweep and downsweep are shaded in figure 4.22.

From (2.27) and (2.28) the frequencies of the low-pass filtered mixed signal in the transponder during the synchronization upsweep f_1^G and downsweep f_2^G are given by:

$$f_1^G = \Delta f + \Delta f_a - \alpha\mu\Delta t + \mu\left(\hat{t} + \Delta t - \frac{T}{2}\right)(\alpha - 1), \tag{4.38}$$

$$f_2^G = \Delta f + \Delta f_a + \alpha\mu\Delta t - \mu\left(\hat{t} + \Delta t - \frac{T}{2}\right)(\alpha - 1). \tag{4.39}$$

Note, that the additional offset in frequency Δf_a of the local signal in the transponder is added to the frequency of the low-pass filtered mixed signal. For the synchronization upsweep μ_1 is replaced by μ and T_1 is substituted by T. For the synchronization downsweep μ_2 is replaced by $(-\mu)$ and T_2 is substituted by T. By definition $\hat{t}$ is an alternative time basis with ($\hat{t} = 0$) at the beginning of the local sweeps in the transponder (2.26). The offset in time is assumed to be constant during the synchronization sweeps for algebraic simplicity here. The Doppler frequency shift is neglected as well. Equations (4.38) and (4.39) simplify to:

$$f_1^G = \Delta f + \Delta f_a - \mu\Delta t + (\alpha - 1)\mu\left(\hat{t} - \frac{T}{2}\right), \tag{4.40}$$

$$f_2^G = \Delta f + \Delta f_a + \mu\Delta t - (\alpha - 1)\mu\left(\hat{t} - \frac{T}{2}\right). \tag{4.41}$$

Depending on the sign of the offset in time f_1^G and f_2^G are defined for an interval of:

$$0 \leq \hat{t} \leq T - \Delta t \qquad\qquad \forall\Delta t \geq 0, \tag{4.42}$$

$$-\Delta t \leq \hat{t} \leq T \qquad\qquad \forall\Delta t < 0. \tag{4.43}$$

The frequency of the low-pass filtered mixed signal, therefore, changes linearly from

$$f_{1,start}^G = \Delta f + \Delta f_a - \mu\Delta t + \begin{cases} (\alpha - 1)\left(-\dfrac{B}{2}\right) & \forall\Delta t \geq 0 \\[3mm] (\alpha - 1)\left(-\dfrac{B}{2} - B\dfrac{\Delta t}{T}\right) & \forall\Delta t < 0 \end{cases} \tag{4.44}$$

to

$$f_{1,stop}^G = \Delta f + \Delta f_a - \mu\Delta t + \begin{cases} (\alpha - 1)\left(\dfrac{B}{2} - B\dfrac{\Delta t}{T}\right) & \forall\Delta t \geq 0 \\[3mm] (\alpha - 1)\left(\dfrac{B}{2}\right) & \forall\Delta t < 0 \end{cases} \tag{4.45}$$

during the synchronization upsweep. Similarly, it changes linearly from

$$f_{2,start}^G = \Delta f + \Delta f_a + \mu\Delta t - \begin{cases} (\alpha - 1)\left(-\dfrac{B}{2}\right) & \forall\Delta t \geq 0 \\[3mm] (\alpha - 1)\left(-\dfrac{B}{2} - B\dfrac{\Delta t}{T}\right) & \forall\Delta t < 0 \end{cases} \tag{4.46}$$

to

$$f_{2,stop}^G = \Delta f + \Delta f_a + \mu \Delta t - \begin{cases} (\alpha - 1)\left(\dfrac{B}{2} - B\dfrac{\Delta t}{T}\right) & \forall \Delta t \geq 0 \\ (\alpha - 1)\left(\dfrac{B}{2}\right) & \forall \Delta t < 0 \end{cases} \tag{4.47}$$

during the synchronization downsweep.

The mismatch of the sweep rates of the radar stations is a direct consequence of the deviation of their respective clock frequencies. The stability of the oscillators which are currently used to clock the radar units is $\pm 25\,\mathrm{ppm}$. Consequently, the deviation of the clock frequencies is within $\pm 50\,\mathrm{ppm}$. Therefore, from (2.35) the ratio α of the sweep rates of the transponder and the base station can be expressed in terms of the deviation δ_{clk} of the clock frequencies. It is given by:

$$\alpha \approx 1 + 2\delta_{clk}. \tag{4.48}$$

It follows from (4.44) and (4.45) that the frequency of the low-pass filtered mixed signal during the synchronization upsweep is a low-bandwidth chirp. The bandwidth of the chirp is:

$$|b_{ts}| = B\left(1 - \frac{|\Delta t|}{T}\right)|\alpha - 1| \tag{4.49}$$

$$\approx 2|\delta_{clk}|B\left(1 - \frac{|\Delta t|}{T}\right), \tag{4.50}$$

and its center frequency is given by:

$$f_{1,center}^G = \Delta f + \Delta f_a - \mu \Delta t - \delta_{clk}B\frac{\Delta t}{T}. \tag{4.51}$$

Here the term "chirp" is used for the linearly changing frequency of the low-pass filtered mixed signal to avoid confusion with the FMCW radar "sweeps" in the 5.8 GHz ISM band.

Similarly to (4.51), the frequency during the synchronization downsweep is obtained from (4.46) and (4.47). It resembles a chirp with the same bandwidth centered at:

$$f_{2,center}^G = \Delta f + \Delta f_a + \mu \Delta t + \delta_{clk}B\frac{\Delta t}{T}. \tag{4.52}$$

The chirp bandwidth $|b_{ts}|$ of the low-pass filtered mixed signal increases linearly with the bandwidth B of the synchronization sweeps and the deviation δ_{clk} of the clock frequencies. In section 4.7.3 it will be shown how the peaks in the power spectral densities widen as the bandwidth or the deviation of the clock frequencies increase.

The center frequency of the chirps of the low-pass filtered mixed signal depends on the deviation of the clock frequencies of the base station and the transponder. The average frequency $f_{1,center}^G$ (4.51) differs from the frequency f_1 (2.43) used for the derivation of the synchronization algorithm by $(-\delta_{clk}B\Delta t/T)$. Similarly, $f_{2,center}^G$ (4.52) differs from

f_2 (2.44) by $(\delta_{clk} B \Delta t / T)$. However, the offset in frequency of both stations is still obtained correctly from (2.46). It is given by:

$$\Delta f = \frac{f^G_{2,center} + f^G_{1,center}}{2} - \Delta f_a. \tag{4.53}$$

If the offset in frequency is known the offset in time can be calculated similarly to (2.45). The offset in time then is:

$$\Delta t = \frac{T}{B} \frac{f^G_{2,center} - f^G_{1,center}}{2} \frac{1}{1 + 2\delta_{clk}}. \tag{4.54}$$

For the system at hand, however, the maximum offset in time prior to synchronization is 400 ns (4.16). Even at the maximum deviation of the clock frequencies of 50 ppm the estimation error of the offset in time is approximately 0.4 ps if the deviation of the clock frequencies is neglected. The round-trip distance $(2d)$, therefore, is estimated with a mean error of less than 0.12 mm which is acceptable for practical application.

Distance Measurement in the Base Station

If the sweep rate of the transponder cannot be adjusted to the sweep rate of the base station during synchronization, then the frequency of the low-pass filtered mixed signal in the base station changes linearly during the measurement sweeps as well. This is illustrated in figure 4.23 where the sweep rate of the received signal $x^\alpha_{bs,rx}$ is larger than the sweep rate of the local signal $x_{bs,lo}$ in the base station. The problem is similar to the effect of the mismatch of the sweep rates during the synchronization sweeps.

Since the estimation errors of the offsets in time and in frequency are negligible, the time dependent general solution for the frequency of the low-pass filtered mixed signal during the measurement upsweep and downsweep, given by:

$$f^G_{up} = \Delta f_{a,up} - \alpha\mu (2t_d) + \mu \left(\hat{t} - \frac{T}{2} \right) (\alpha - 1), \tag{4.55}$$

$$f^G_{dn} = \Delta f_{a,dn} + \alpha\mu (2t_d) - \mu \left(\hat{t} - \frac{T}{2} \right) (\alpha - 1), \tag{4.56}$$

is obtained from (2.78) by substituting μ_{meas} by μ for the measurement upsweep and by $(-\mu)$ for the measurement downsweep. The sweep duration $T_{meas} = T$ is identical for both sweeps. The equations are valid for $(2t_d \le \hat{t} \le T)$ where $\hat{t} = 0$ is assumed at the beginning of the local sweeps $x_{bs,lo}$ when the sampling of the low-pass filtered mixed signal starts in the base station.

The frequency of the low-pass filtered mixed signal, therefore, changes linearly from

$$f^G_{up,start} = \Delta f_{a,up} - 2t_d\mu - \frac{B}{2} (\alpha - 1) \tag{4.57}$$

to

$$f^G_{up,stop} = \Delta f_{a,up} - 2t_d\alpha\mu + \frac{B}{2} (\alpha - 1) \tag{4.58}$$

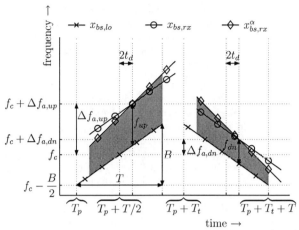

Figure 4.23: Distance measurement, different sweep rates (base station): If the sweep rates of the received and the locally generated signal do not match, the frequency of the low-pass filtered mixed signal in the base station is not constant during the measurement upsweep or downsweep (shaded areas). This can be modeled as a chirp around the frequencies f_{up} and f_{dn}. Here, the sweep rate of the transponder is larger than the sweep rate of the base station.

during the measurement upsweep. Similarly, it changes linearly from

$$f^G_{dn,start} = \Delta f_{a,dn} + 2t_d\mu + \frac{B}{2}(\alpha - 1) \tag{4.59}$$

to

$$f^G_{dn,stop} = \Delta f_{a,dn} + 2t_d\alpha\mu - \frac{B}{2}(\alpha - 1) \tag{4.60}$$

during the measurement downsweep.

Next, the ratio α of the sweep rates of the transponder and the base station is expressed by the deviation δ_{clk} of the respective clock frequencies (4.48). The frequency of the low-pass filtered mixed signal during the measurement upsweep and downsweep, again, describes low-bandwidth chirps. Their bandwidth is given by:

$$|b_{bs}| = 2|\delta_{clk}| B\left(1 - \frac{2t_d}{T}\right), \tag{4.61}$$

and the respective center frequencies of the chirps are:

$$f^G_{up,center} = \Delta f_{a,up} - \frac{2d}{c_{ph}}\mu(1 + \delta_{clk}), \tag{4.62}$$

$$f^G_{dn,center} = \Delta f_{a,dn} + \frac{2d}{c_{ph}}\mu(1 + \delta_{clk}). \tag{4.63}$$

Consequently, the peaks in the power spectral densities are shifted and widened as the deviation of the clock frequencies of the radar stations increases.

If the average frequencies of the low-pass filtered mixed signal $f^G_{up,center}$ and $f^G_{dn,center}$ are applied to (2.88) the distance measured between the radar stations is given by:

$$d_{meas} = d\left(1 + \delta_{clk}\right). \tag{4.64}$$

The measurement error with respect to the true distance d is linear in d and the deviation δ_{clk} of the clock frequencies of the transponder and the base station.

For the hardware setup presented in chapter 3 the stability of the oscillators is ± 25 ppm and the maximum deviation of the clock frequencies of two radar stations is ± 50 ppm. Therefore, the maximum error of the distance measurements due to the mismatch of the sweep rates theoretically does not exceed ± 50 ppm of the true distance. For short distances it can easily be neglected. However, at large distances a measurement error of 50 ppm, i.e. 10 cm at a distance of 2 km, is acceptable too.

Firstly, the standard deviation of the measurement results increases significantly for long distances due to the low signal level of the received signal and the resulting low SNR of the low-pass filtered mixed signal. Furthermore, estimation errors occur since the shape of the peaks in the power spectral densities deviates from a parabola as the deviation of the clock frequencies increases and the parabolic interpolation becomes more inaccurate. Therefore, the theoretical increase of the measured distance by up to 50 ppm is not the dominant source of error. In the following it is shown how the shape of the peaks in the power spectral densities changes as the deviation of the clock frequencies of the radar stations is increased.

4.7.2 Window Functions Revised

The sweep rate of the transponder cannot be adjusted to the sweep rate of the base station with the current hardware setup. Therefore, the frequency of the low-pass filtered mixed signal in the transponder and the base station changes linearly during the synchronization and measurement sweeps, if the clock frequencies of the stations do not match. The bandwidth of the chirps of the low-pass filtered mixed signal in both stations then is given by (4.50) and (4.61), respectively. Both equations simplify to:

$$|b_{ts}| \approx |b_{bs}| \approx 2\,|\delta_{clk}|\,B, \tag{4.65}$$

since the sweep duration $T \approx 0.9845$ ms is much larger than the maximum offset in time $\Delta t \leq 400$ ns (4.16) and the round-trip time-of-flight corresponding to the maximum allowable distance of the radar stations $2t_{d_{max}} \leq 14.9\,\mu$s (4.27) in practice.

As the chirp bandwidth increases the side lobe attenuation of the window function applied to the sampled data becomes more important. In figure 4.24 the spectra obtained with different windows are compared. The frequency deviation of the stations is set to 40 ppm and a rectangular window (gray line) and a Blackman window (black line) are applied to the same set of samples.

Clearly, the latter outperforms the rectangular window. Without an additional window function there is no unique peak in the power spectral density. In fact, the peak

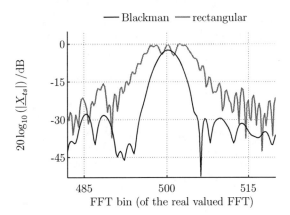

Figure 4.24: Widening of the peaks and window functions (simulated data): For large deviations of the clock frequencies of the stations, the selection of the window function is critical to the performance of the measurement system. Here, a Blackman window and a rectangular window are applied to the low-pass filtered mixed signal at a clock deviation of 40 ppm. There is no unique peak in the power spectral density if the rectangular window is used.

magnitude of the gray line at bin 502 is off the "true" position of the peak by about 2 FFT bins. Hence, a window with a small side lobe level, e. g. the Blackman window, must be applied to the data especially for large frequency deviations.

Figure 4.24 can also be compared to figure 4.5 where identical sweep rates have been assumed for both radar stations. In figure 4.5, clearly, the width of the peak obtained with a Blackman window is larger than the width of the peak obtained with the rectangular window. However, in section 4.2.4 it has been shown that the algorithm for the estimation of the frequency of the low-pass filtered mixed signal yields more accurate results if the Blackman window is used, even if the sweep rates of the radar stations match.

Therefore, a Blackman window is applied to sampled data regardless of the deviation of the clock frequencies. In the following it is shown how the shape of the peaks in the power spectral densities changes as the deviation of the clock frequencies increases.

4.7.3 Widening of the Peaks

Even if a Blackman window is applied to the low-pass filtered mixed signal the spectra of the chirps obtained from the FFT differ from the spectra of an ideal sinusoidal signal significantly. Figure 4.25 depicts spectra obtained from measured and simulated data for deviations of the clock frequencies of the base stations and the transponder of 0 ppm, 20 ppm, and 40 ppm. For the simulations and measurements the bandwidth of

Table 4.3: Average peak width ($-6\,$dB) and peak attenuation (measured data): As the absolute deviation $|\delta_{clk}|$ of the clock frequencies of both stations increases, the peak width increases as well. Furthermore, the signal level decreases relative to the level at a clock frequency deviation of $0\,$ppm.

deviation of clock frequencies in ppm	peak width ($-6\,$dB) mean in Hz	std in Hz	peak attenuation mean in dB	std in dB
0	2362	3	0	-
10	2479	19	-0.35	0.04
20	3026	45	-1.21	0.07
30	3985	78	-2.36	0.05
40	4770	84	-3.17	0.06
50	6090	108	-4.18	0.10

the FMCW radar signals is set to $132\,$MHz. This matches the parameters of the system setup given in table 3.1.

In the simulations the chirp bandwidth is calculated from (4.65). The spectra depicted in figure 4.25(a) are obtained from simulated data. They match the spectra obtained from measured data well, which are shown in figure 4.25(b). As the magnitude of the deviation of the clock frequencies increases the width of the peaks in the power spectral densities increases as well. Coincidentally, the peak magnitude in the power spectral densities decreases while the noise level does not change significantly.

Table 4.3 summarizes the $-6\,$dB width of the peaks and the respective attenuation of the maxima measured for deviations of the clock frequencies from $0\,$ppm to $50\,$ppm. As the frequency deviation increases the width of the peaks increases from $2.4\,$kHz to $6.1\,$kHz. The magnitude of the peak is decreased by up to $4.2\,$dB. The smallest peak width is obtained if the clock frequencies of both stations match exactly. The peak width is then close to the theoretical width of the Blackman window of $2.35\,$FFT bins [35]. This indicates a constant frequency of the low-pass filtered mixed signal during the entire sampling time and hence an excellent linearity of the sweeps of both radar stations.

For large deviations of the clock frequencies of the radar stations, the respective mismatch of the sweep rates severely affects the performance of the measurement system. This is shown in the next section.

4.7.4 Accuracy of the Synchronization and Measurement Results

The algorithm for the estimation of the frequency of the low-pass filtered mixed signal presented in section 4.2 utilizes a parabolic interpolation of the peak position to overcome the limited resolution of the FFT. According to Smith, the main lobe in the power spectral density of a single sinusoidal signal can be replaced by a quadratic polynomial for any practical window transform in a sufficiently small neighborhood about the peak [86]. However, as shown in figure 4.25 the shape of the peak changes as the mismatch of

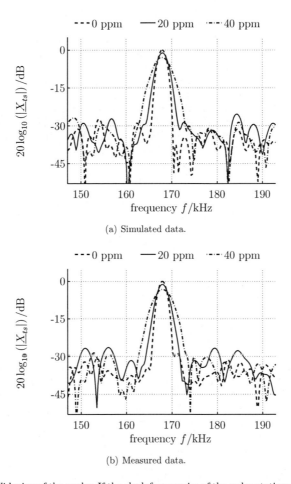

(a) Simulated data.

(b) Measured data.

Figure 4.25: Widening of the peaks: If the clock frequencies of the radar stations do not match, the sweep rate of the transponder differs from the sweep rate of the base station. The width of the peaks in the power spectral densities increases as the absolute deviation $|\delta_{clk}|$ of the clock frequencies of both stations, and hence the mismatch of their respective sweep rates, increases. Coincidentally, the level of the peaks decreases.

the sweep rates increases. Consequently, the approximation of the peak by a quadratic polynomial becomes more inaccurate as the mismatch of the sweep rates increases.

Furthermore, the peak magnitude in the power spectral densities decreases by up to 4.2 dB as the deviation of the sweep rates increases. Consequently, the effective SNR of the low-pass filtered mixed signal is reduced. It has already been shown in section 4.5.5 that the standard deviation of the estimation errors during synchronization and measurement increases as the SNR of the low-pass filtered mixed signal decreases.

For the simulation results presented in figure 4.26 and figure 4.27 an SNR of the low-pass filtered mixed signal of 20 dB is assumed. This corresponds to the maximum SNR which can be achieved with the current hardware setup due to phase noise of the PLL. Furthermore, a bandwidth of the FMCW sweeps of 132 MHz is assumed. Since the frequency stability of the oscillator of each LPR unit is ±25 ppm, the deviation of the clock frequencies of the base station and the transponder is in:

$$-50\,\text{ppm} \leq \delta_{clk} \leq 50\,\text{ppm}. \tag{4.66}$$

Consequently, the chirp bandwidth of low-pass filtered mixed signal (4.65) changes from 0 Hz if the clock frequencies of both stations match to approximately 13.2 kHz at a frequency deviation of ±50 ppm.

The accuracy of the estimation of the offsets in time Δt_1 and in frequency Δf of the base station and the transponder strongly depends on the mismatch of the sweep rates of both stations and hence on the deviation of their clock frequencies. The standard deviation of the respective estimation errors, δt_1 and δf, is shown in figure 4.26. If the sweep rates of the radar stations match a minimum standard deviation of the estimation errors of 18 ps and 2.5 Hz is obtained. However, as the mismatch of the sweep rates increases the width of the peaks in the power spectra increases as well. In the worst case setup the deviation of the clock frequencies is ±50 ppm. The offsets in time and in frequency are then estimated with a standard deviation of 160 ps and 22 Hz, respectively.

Similarly, the standard deviation of the distance measurements increases as the mismatch of the sweep rates increases. It is depicted in figure 4.27 for deviations of the clock frequencies from −50 ppm to 50 ppm. The mismatch of the sweep rates is at its maximum at a deviation of the clock frequencies of ±50 ppm. The standard deviation of the distance measurements is then 30 mm. If the clock frequencies, and hence the sweep rates, of the stations match, a standard deviation of the distance measurements of 6.3 mm is obtained if the clock jitter of the sweep triggers described in section 4.4.5 is considered (black line).

If the clock jitter is neglected, i. e. if ideal electronic components are assumed, the standard deviation of the distance measurements is as low as 3.6 mm (gray line). Therefore, it can be concluded that the accuracy of the measurement results for small deviations of the clock frequencies of the radar stations is limited by the phase noise of the PLL in the signal generator and the clock jitter. If the clock frequencies differ by more than ±30 ppm the mismatch of the sweep rates becomes the dominant source of error.

In the simulations presented here an SNR of the low-pass filtered mixed signal of 20 dB has been assumed. For long distances, however, the effective SNR is even less than 20 dB. The performance of the measurement system then degrades according to

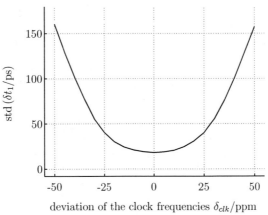

(a) Estimation error δt_1 of the offset in time.

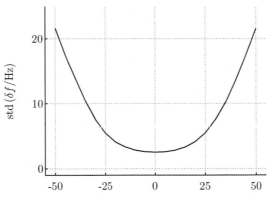

(b) Estimation error δf of the offset in frequency.

Figure 4.26: Dependence of the synchronization results on the deviation of the clock frequencies of the radar stations (simulated data, transponder): The standard deviation of the estimation error of the offsets in time δt_1 and in frequency δf depends on the deviation of the clock frequencies of the base station and the transponder. Both offsets are estimated with a minimum standard deviation of 18 ps and 2.5 Hz, respectively, if the clock frequencies of the radar stations match. If the deviation of the clock frequencies is ±50 ppm the standard deviation of both estimation errors is 160 ps and 22 Hz, respectively.

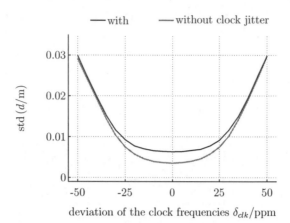

Figure 4.27: Dependence of the measurement results on the deviation of the clock frequencies
(simulated data, base station): The standard deviation of the distance measurements depends on the deviation of the clock frequencies of the radar stations.
The distance is estimated with a standard deviation of 30 mm if the deviation of
the clock frequencies is ±50 ppm. The standard deviation is 6.3 mm if the clock
frequencies of the stations match and the clock jitter of the sweep triggers is considered (black line). It is as low as 3.6 mm if the clock jitter is neglected (gray
line).

figure 4.18 and figure 4.19 and the loss in SNR due to the mismatch of the sweep
rates impairs the accuracy of the synchronization and measurement results even more
severely.

During the calculation of the offset in time, the offset in frequency, and the distance
of the radar stations some system parameters, e. g. the sampling frequency, are assumed to be constant. However, they can deviate from their nominal values in practice.
The errors resulting from the inaccurate knowledge of constant system parameters are
treated in the next section.

4.8 Deviation of Constant System Parameters from Their Nominal Values

The clock frequency of the radar units and the phase velocity of the radar signals are
two important system parameters which are assumed to be known constants during the
synchronization and measurement process. If they deviate from their nominal values,
however, an error is introduced to the synchronization and measurement results.

4.8.1 Clock Frequency

The offsets in time and in frequency are calculated from the frequency of the low-pass filtered mixed signal in the transponder during synchronization. Similarly, the distance of the radar stations is calculated from the frequency of the low-pass filtered mixed signal in the base station. The sampling frequency the low-pass filtered mixed signal is digitized with, however, is derived from the clock frequency of each radar module. Since the frequency stability of the oscillators used to clock the radar units is $\pm 25\,$ppm, the sampling frequency can deviate from its nominal value by the same proportion.

Clock Frequency in the Transponder

The offsets in time and in frequency of the base station and the transponder are calculated from the frequencies of the low-pass filtered mixed signal in the transponder during the synchronization upsweep f_1 and downsweep f_2. The offsets in time and in frequency are given by (2.60) and (2.59), respectively.

The accuracy of the frequency estimation depends on the absolute value of the sampling frequency of the transponder. During the spectral analysis the peak position is calculated in FFT bins at first. Ideally, the peak position is given by:

$$m_{z,ip} = \frac{f_{ts}}{f_s} N, \tag{4.67}$$

where f_{ts} is the frequency of the low-pass filtered mixed signal, f_s is the sampling frequency, and N is the number of samples acquired. However, since f_s is derived from the clock frequency of the station by various prescalers, it can deviate from its nominal value by up to $\pm 25\,$ppm. As the true sampling frequency increases the position $m_{z,ip}$ corresponding to a constant frequency f_{ts} decreases.

However, during synchronization the frequency of the low-pass filtered mixed signal is calculated from the position of the peak in the power spectral density. The nominal value of the sampling frequency is used for the conversion since there is no way to estimate the true sampling frequency. Clearly, an error occurs if the true sampling frequency does not match its nominal value. If the true sampling frequency is larger than its nominal value then $m_{z,ip}$ is too small and the frequency of the low-pass filtered mixed signal is estimated too small. If the true sampling frequency is smaller than its nominal value, $m_{z,ip}$ is too large and the frequency of the low-pass filtered mixed signal is estimated too large.

Consequently, the offsets in time and in frequency are estimated with an error as well. The estimation error is proportional to the true offsets in time and in frequency. It depends on the deviation of the true sampling frequency from its nominal value as well. Therefore, the initial offsets in time and in frequency have to be kept as low as possible for high precision synchronization. For the measurement system presented here the maximum offsets in time and in frequency prior to synchronization are $400\,$ns (4.16) and $290\,$kHz (3.17), respectively. They are estimated with a maximum error of $0.1\,$ps and $7.25\,$Hz, respectively, if the sampling frequency in the transponder deviates from its nominal value by $25\,$ppm.

If necessary the synchronization can be repeated. The initial offsets in time and in frequency of each synchronization step are then given by the synchronization error of the previous step, and the residual error due to the inaccurate sampling frequency can be reduced iteratively. However, even if the synchronization is not repeated the maximum estimation error of the offset in frequency is close to the resolution of the DDS in the current hardware setup. The maximum estimation error of the offset in time corresponds to an estimation error of the round-trip distance $(2d)$ of 0.03 mm. Therefore, the synchronization errors due to the inaccurate sampling frequency in the transponder can be neglected for practical application.

Clock Frequency in the Base Station

A similar systematic error arises if the clock frequency of the base station does not match its nominal value. For the hardware setup presented in chapter 3 the true clock frequency, given by:

$$f_{clk,bs} = (1 + \delta_{clk,bs})\, f_{clk}, \tag{4.68}$$

can differ from its nominal value f_{clk} by $-25\,\text{ppm} \leq \delta_{clk,bs} \leq 25\,\text{ppm}$.

Hence, the sweep rate of the sweeps of the base station, which is:

$$\mu_{bs} = \mu\,(1 + \delta_{clk,bs})^2\,, \tag{4.69}$$

differs from its nominal value μ as well (2.32), (2.33), (2.35).

During synchronization the transponder estimates its offsets in time Δt and in frequency Δf relative to the base station. Furthermore, the mismatch of the sweep rates of both stations α is calculated from Δf (2.35). It has already been shown that the transponder cannot adjust its sweep rate to the sweep rate of the base station with the current hardware setup. Then, the mismatch of the sweep rates is the dominant source of error.

If the hardware setup is modified, e. g. if a VCXO or high resolution DDS is used with the system, the transponder can adjust its sweep rate. Similarly to (2.83) and (2.84), the true frequency of the low-pass filtered mixed signal in the base station during the measurement sweeps is then given by:

$$f_{up} = \Delta f_{a,up}\,(1 + \delta_{clk,bs}) - 2\frac{d}{c_{ph}}\mu_{bs}\,(1 + \delta_{clk,bs})^2\,, \tag{4.70}$$

$$f_{dn} = \Delta f_{a,dn}\,(1 + \delta_{clk,bs}) + 2\frac{d}{c_{ph}}\mu_{bs}\,(1 + \delta_{clk,bs})^2\,, \tag{4.71}$$

where $\Delta f_{a,up}$ and $\Delta f_{a,dn}$ are the additional offsets in frequency of the synchronized reply of the transponder, d is the true distance of the stations, and c_{ph} is the phase velocity of the radar signals.

The low-pass filtered mixed signal is sampled with a frequency of $(1 + \delta_{clk,bs})\, f_s$ during both measurement sweeps. However, the nominal value f_s is used to convert the positions of the maxima in the power spectral densities to the frequencies of the signal.

Consequently, the frequencies of the low-pass filtered mixed signal in the base station during the measurement upsweep and downsweep are computed as:

$$f_{up}^{calc} = \Delta f_{a,up} - 2\frac{d}{c_{ph}}\mu_{bs}\left(1 + \delta_{clk,bs}\right), \tag{4.72}$$

$$f_{dn}^{calc} = \Delta f_{a,dn} + 2\frac{d}{c_{ph}}\mu_{bs}\left(1 + \delta_{clk,bs}\right). \tag{4.73}$$

Similarly to (2.88), the distance is calculated subsequently from f_{up}^{calc} and f_{dn}^{calc}. It is found to be:

$$d_{meas} = d\left(1 + \delta_{clk,bs}\right). \tag{4.74}$$

Clearly, the measured distance d_{meas} differs from the true distance d of the radar stations if the clock frequency of the base station does not match its nominal value. The measurement error is linear in the distance d and the deviation $\delta_{clk,bs}$ of the clock frequency of the base station from its nominal value. Similar results can be derived for the relative velocity.

To demonstrate the effect of the clock frequency of the base station on the distance measurements the base station is clocked by a signal generator. The transponder is equipped with a VCXO. The clock frequency of the base station is now changed within the pull range of the VCXO. If an offset in frequency is detected during synchronization the transponder automatically adjusts its clock frequency. Thus, the clock frequency and hence the sweep rate of the transponder match the parameters of the base station over the entire pull range of the VCXO, and the effect of the deviation of the clock frequency of the base station from its nominal value can be measured. The results are shown in figure 4.28. The measured distance and hence the distance measurement error increase linearly as the deviation of the clock frequency of the base station from its nominal value increases.

From (4.74) the distance measurement error is linear in the deviation of the clock frequency of the base station from its nominal value. Therefore, oscillators with a low frequency tolerance should be used for high precision measurements. For the hardware setup presented in chapter 3 the clock frequency of the base station deviates from its nominal value by at most 25 ppm. This corresponds to an average measurement error of 5 cm at a distance of 2 km. If this error is not acceptable and a smaller inaccuracy is required the measurement system has to be calibrated at a large distance. Since the error is linear in the deviation of the clock frequency and the distance of the stations, a constant factor can be used in (2.88) to compensate for the deviation of the clock frequency. However, it is doubtful that a distance of 2 km can easily be referenced with a smaller inaccuracy.

The distance is also measured with an error if the phase velocity of the radar signals is not known exactly. This is shown in the following.

4.8.2 Phase Velocity

The local positioning radar LPR presented in this thesis calculates the distance of two radar units from the round-trip time-of-flight of a signal. The phase velocity of the

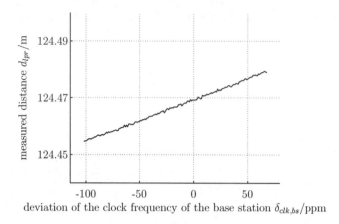

Figure 4.28: Deviation of the clock frequency of the base station from its nominal value (measured data, base station): The base station is clocked by a signal generator and the transponder is equipped with a VCXO. The clock frequency of the base station is changed within the pull range of the VCXO, such that the transponder can adjust its clock frequency to the clock frequency of the base station. Clearly, the measured distance increases linearly with the deviation of the clock frequency of the base station from its nominal value.

radar signals is assumed to be a constant value which is used to convert the RTOF into the distance. In practice, the phase velocity of the radar signals is slightly smaller than the vacuum speed of light c_0. Its exact value, which is given by:

$$c_{ph} = \frac{c_0}{\sqrt{\epsilon_r}}, \qquad (4.75)$$

is related directly to the relative permittivity or dielectric constant ϵ_r of the medium the electromagnetic waves propagate through [85]. The relative permittivity of air depends on environmental conditions like the atmospheric pressure, the relative humidity, and the temperature, and on the exact wavelength of the radar signals. Note, that a relative magnetic permeability of 1 has been assumed in (4.75) [85].

For the measurements presented in figure 4.29 the phase velocity of the radar signals is assumed to be the vacuum speed of light, $c_{ph} = c_0 = 299792458 \, \text{m/s}$. The results are compared to the results of a laser ranging system [49]. If the phase velocity of the FMCW radar signals was known exactly the difference between the measurement results of the LPR and the laser ranging system would remain constant. Since the true phase velocity of the radar signals is smaller than the vacuum speed of light the distance is measured too large by the LPR. Hence, the measurement error of the LPR with respect to the laser ranging system increases proportionally to the distance of the radar stations. The offset of approximately 11.8 m is caused by the cables used to connect the radar stations to their antennas.

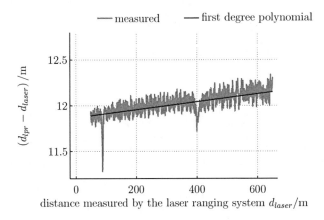

Figure 4.29: Phase velocity of the signals (measured data, base station): The distance between both LPR units is increased from 50 m to 650 m. It is compared to the distance measured by a laser ranging system. The phase velocity of the radar signals is assumed to be $c_0 = 299792458$ m/s. However, the phase velocity of the signals depends on atmospheric conditions and does not match the speed of light in vacuum exactly. Therefore, the difference of the distances measured by the LPR d_{lpr} and by the laser ranging system d_{laser} increases as the distance between the stations increases. The sudden decay in the measured distance at distances of 88 m and 400 m is caused by ground echoes.

The sudden decay in the measured distance at distances of 88 m and 400 m is caused by interference of multipath components due to reflections on the ground. Ground echoes can be avoided by mounting the antennas at a higher clearance above the ground. In the setup used for the measurements in figure 4.29, however, the antennas were mounted at a height of 1.4 m only.

Measurements of the relative permittivity of air at a wavelength of 5.2 cm, which corresponds to the frequency of the radar signals of 5.8 GHz, are not available. However, measurements at similar wavelengths are reported by Crain [15] and Phillips [64]. Crain estimates the relative permittivity of dry air at a wavelength of 3.2 cm to be 1.000572. A similar dielectric constant of 1.0005548 at a wavelengh of 10 cm is reported by Phillips. Therefore, it is assumed in the following that the relative permittivity of air at a wavelength of 5.2 cm is close to the results reported by Crain and Phillips.

Phillips observes the dependence of the relative permittivity of air on the atmospheric pressure and the relative humidity [64]. The results are shown in table 4.4 and table 4.5. Both parameters affect the relative permittivity significantly. The relative permittivity changes from 1.000129 to 1.000555 as the atmospheric pressure increases from 239.5 hPa to 1009.6 hPa. At an atmospheric pressure of 1000 hPa it changes from 1.000555 for dry air to 1.000806 for air saturated with water vapor.

Table 4.4: Relative permittivity, atmospheric pressure: The relative permittivity ϵ_r of dry air increases as the atmospheric pressure increases [64].

relative permittivity ϵ_r	atmospheric pressure in hPa	temperature in °C
1.000555	1009.6	25.5
1.000393	738.8	22.0
1.000278	510.7	23.0
1.000129	239.5	22.0

Table 4.5: Relative permittivity, relative humidity: The relative permittivity ϵ_r increases as the relative humidity increases [64].

relative permittivity	atmospheric pressure in hPa	temperature in °C	relative humidity in %
1.000806	1000.8	22.0	saturated
1.000735	1000.7	21.3	88.1
1.000728	1000.6	22.0	83.3
1.000687	1004.5	19.5	63.1
1.000668	1006.1	19.0	55.7
1.000555	1009.6	25.5	dry

The measurements presented in figure 4.29 are repeated at an atmospheric pressure of approximately 1010 hPa and a relative humidity of approximately 83 %. From table 4.5 a relative permittivity of $\epsilon_r = 1.000073$ is assumed for the measurements. Consequently the phase velocity is assumed to be:

$$c_{ph} = 299683074 \, \frac{\text{m}}{\text{s}}. \tag{4.76}$$

The average difference of the distance measurements obtained by the LPR and the laser system then does not change significantly over distance anymore. Therefore, the phase velocity given by (4.76) is used for the measurements presented in chapter 5 as well.

In practice, the phase velocity of the radar signals is usually not known exactly. Consequently, a measurement error occurs if (2.88) is used to calculate the distance of the radar stations. The error is linear in the deviation of the phase velocity from its nominal value. In a typical outdoor environment the phase velocity of the radar signals, given by (4.75), changes from $(c_0/1.0002)$ for dry air and an atmospheric pressure of 740 hPa to $(c_0/1.0005)$ for saturated air and an atmospheric pressure of 1050 hPa. Consequently, the distance measurement results change by approximately ± 150 ppm over environmental conditions, e. g. ± 30 cm at a distance of 2 km. If a smaller inaccuracy is required the system must be calibrated at a known distance. A large distance should be used for the calibration since the error is linear in the measured distance as well.

In the previous sections all potential errors caused by the hardware setup and the configuration of the LPR have been treated in detail. However, the effects of the environment the system is used in remain to be discussed. Especially, in indoor environments multipath propagation can limit the performance of the system. In the next section it is shown how multipath distortions can be mitigated.

4.9 Multipath Propagation

Up to now a single line-of-sight (LOS) transmission from the base station to the transponder and back has been assumed. However, the signal is transmitted through a wireless channel. Consequently, multiple copies of the transmitted signal arrive at the receiving station through non-line-of-sight (NLOS) propagation paths due to reflection, scattering, and refraction [67,83]. The interference of the NLOS signal components and the signal received through the LOS path can result in synchronization and measurement errors.

4.9.1 Mitigation of Multipath Distortions

The mitigation of multipath distortions during the synchronization sweeps in the transponder and during the measurement sweeps in the base station is treated in detail in the following.

Synchronization in the Transponder

For algebraic simplicity (2.43) and (2.44) are used to demonstrate the effect of multipath propagation on the frequency of the low-pass filtered mixed signal in the transponder during synchronization. Changes in the offset in time due to the deviation of the clock frequencies of the base station and the transponder are not taken into account. The Doppler frequency shift is neglected as well.

In a simple model for multipath signal propagation through physical channels, the received signal in the transponder $x_{ts,rx}$ is a sum of $(I + 1)$ attenuated and delayed copies of the transmitted signal $x_{bs,tx}$ [65], i.e.:

$$x_{ts,rx}(t) = \sum_{i=0}^{I} \left(a_i x_{bs,tx}(t - \tau_i)\right). \qquad (4.77)$$

The delays of the signal components with respect to the time the signal $x_{bs,tx}$ is transmitted by the base station are denoted by:

$$\tau_0 < \tau_1 < \cdots < \tau_i < \cdots < \tau_I. \qquad (4.78)$$

For the LOS path the delay is proportional to the distance d between the stations. It is given by:

$$\tau_0 = t_d. \qquad (4.79)$$

The delays τ_i $(i > 0)$ of the NLOS signal components are larger than τ_0 due to the longer time-of-flight. The attenuation of the multipath components is neglected in the

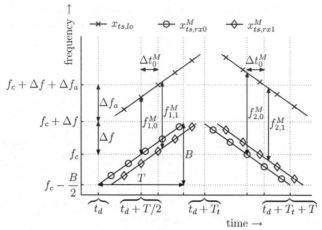

Figure 4.30: Synchronization, multipath environment (transponder): In a multipath environment multiple copies of the transmitted signal arrive at the transponder. The component of the received signal corresponding to the LOS path $x_{ts,rx0}^M$ has to be used to estimate the offsets in time Δt_0^M and in frequency Δf, i. e. $f_{1,0}^M$, $f_{2,0}^M$ have to be used for the calculations.

following, $a_i = 1$, since the algorithm for synchronization focuses on the frequency of the low-pass filtered mixed signal.

Each of the received copies has a different offset in time Δt_i^M with respect to the local signal in the transponder. By definition the offset in time is positive if the received signal precedes the locally generated signal. Consequently, the signal component corresponding to the LOS path ($i = 0$) between the stations has the largest offset in time. The offsets in time of the multipath components decrease as the delays τ_i increase, i. e.:

$$\Delta t_0^M > \Delta t_1^M > \cdots > \Delta t_i^M > \cdots > \Delta t_I^M. \tag{4.80}$$

The offset in frequency is not changed by multipath propagation and, therefore, is constant for all i.

Figure 4.30 depicts the signal constellation in the transponder during synchronization for $I = 1$. The received signal is the sum a LOS component $x_{ts,rx0}^M$ and an NLOS component $x_{ts,rx1}^M$.

As before the received signal and the local signal $x_{ts,lo}$ are multiplied in the transponder. Similarly to (2.43) and (2.44), the low-pass filtered mixed signal then is a sum of $(I + 1)$ sinusoidal signals. The frequency components are given by:

$$f_{1,i}^M = \Delta f + \Delta f_a - \mu \Delta t_i^M, \tag{4.81}$$
$$f_{2,i}^M = \Delta f + \Delta f_a + \mu \Delta t_i^M, \tag{4.82}$$

where $f_{1,i}^M$ and $f_{2,i}^M$ denote the frequency components of the low-pass filtered mixed signal in the transponder corresponding to the ith multipath during the synchronization upsweep and downsweep, respectively.

Adjacent frequency components can be separated from each other in the power spectral densities of the low-pass filtered mixed signal if their delay $(\tau_{i+1} - \tau_i)$ exceeds a threshold, which is referred to as the axial resolution of the radar system. It is investigated in detail in section 4.9.2. For now it is assumed that all frequency components are well separated from each other in the power spectral densities.

If all frequency components are well separated, then $(I + 1)$ peaks are detected in the power spectra of the low-pass filtered mixed signal in the transponder. The frequency components corresponding to the LOS connection between the base station and the transponder have to be used to estimate the offsets in time Δt_0^M and in frequency Δf of both stations. Equations (4.80), (4.81), and (4.82) imply:

$$f_{1,i}^M \; < \; f_{1,l}^M \qquad \forall i,l : 0 \leq i < l \leq I, \tag{4.83}$$

$$f_{2,i}^M \; > \; f_{2,l}^M \qquad \forall i,l : 0 \leq i < l \leq I. \tag{4.84}$$

Consequently, during synchronization the peak with the lowest frequency $f_{1,0}^M$ is selected when evaluating the upsweep since only the frequency component corresponding to the LOS path $(i = 0)$ is used for synchronization. When evaluating the downsweep, the peak with the highest frequency $f_{2,0}^M$ is selected.

Exemplary spectra of the low-pass filtered mixed signal in the transponder in a multipath environment are shown in figure 4.31. Figure 4.31(a) represents a spectrum recorded during the synchronization upsweep, a spectrum recorded during the synchronization downsweep is depicted in figure 4.31(b). The frequency components $f_{1,0}^M$ and $f_{2,0}^M$ corresponding to the LOS path have been labeled. In figure 4.31 they coincide with the highest peaks in the power spectral densities.

However, there is no guarantee that the LOS signal is the strongest of the arriving signals [63]. Therefore, the selection of the peaks must not be based on the signal level alone. Again, the peak with the lowest frequency $f_{1,0}^M$ has to be selected during the synchronization upsweep and the peak with the highest frequency $f_{2,0}^M$ has to be selected during the synchronization downsweep.

If the frequencies $f_{1,0}^M$ and $f_{2,0}^M$ are used to estimate the offset in time Δt_0^M and the offset in frequency Δf of the base station and the transponder there is no systematic error due to multipath propagation. The synchronization is as accurate as in the case of a single LOS connection between both stations.

Distance Measurement in the Base Station

The measurement sweeps of the transponder are transmitted through the wireless channel as well. The channel is assumed to be time-invariant and reciprocal to the channel during synchronization. Therefore, the signal sent from the transponder to the base station is subjected to multipath propagation as well. Similarly to (4.77), the signal

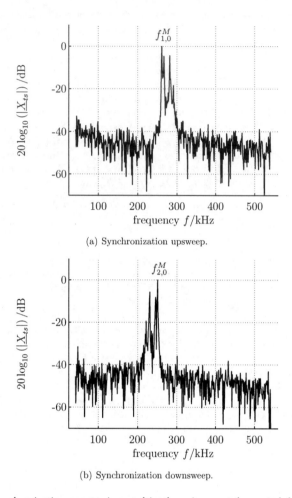

(a) Synchronization upsweep.

(b) Synchronization downsweep.

Figure 4.31: Synchronization, spectra in a multipath environment (measured data, transponder): In a multipath environment multiple peaks are found in the spectrum of the low-pass filtered mixed signal during synchronization. The frequencies corresponding to the LOS path have to be used to calculate the offset in time Δt and in frequency Δf. The peak selection is not based on the signal level but on the frequencies of the peaks. The peaks with the lowest frequency $f_{1,0}^M$ during the synchronization upsweep and the highest frequency $f_{2,0}^M$ during the synchronization downsweep correspond to the LOS path.

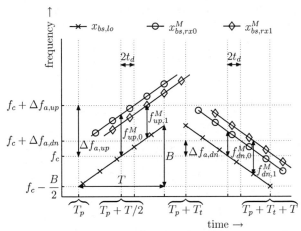

Figure 4.32: Distance and velocity measurement, multipath environment (base station): In a multipath environment multiple copies of the transmitted response arrive at the base station. The component of the received signal corresponding to the LOS path $x_{ts,rx0}^M$ has to be used to estimate the distance and relative velocity of both stations, i. e. $f_{up,0}^M$, $f_{dn,0}^M$ have to be used for the calculations.

$x_{bs,rx}$ received by the base station is a sum of $(I+1)$ delayed and attenuated copies of the signal $x_{ts,tx}$ transmitted by the transponder. It can be written as:

$$x_{bs,rx}(t) = \sum_{i=0}^{I} \left(a_i x_{ts,tr}(t - \tau_i) \right). \tag{4.85}$$

As before the focus is on the frequency of the low-pass filtered mixed signal and the attenuation of the multipath components is neglected, $a_i = 1$.

Figure 4.32 depicts the frequencies of the locally generated signal $x_{bs,lo}$ and the received signal in the base station for one LOS path $x_{bs,rx0}^M$ and one NLOS path $x_{bs,rx1}^M$. The LOS signal component $x_{bs,rx0}^M$ is delayed by the round-trip time-of-flight $2t_d$ with respect to the local sweeps. All NLOS signal components arrive after the LOS signal due to the longer time-of-flight.

The received signal is multiplied with the local signal. Similarly to (2.83) and (2.84), the low-pass filtered mixed signal then is a sum of $(I+1)$ sinusoidal signals. The frequency components are given by:

$$f_{up,i}^M = \Delta f_{a,up} - (2t_d + (\tau_i - \tau_0)) \mu, \tag{4.86}$$
$$f_{dn,i}^M = \Delta f_{a,dn} + (2t_d + (\tau_i - \tau_0)) \mu, \tag{4.87}$$

where $i = 0$ describes the LOS propagation path and $(\tau_i - \tau_0)$ is the delay of the ith NLOS signal with respect to the LOS signal.

From (4.78) the frequency of the low-pass filtered mixed signal during the measurement upsweep decreases as the index i and hence the delay τ_i increases, i. e.:

$$f_{up,i}^M > f_{up,l}^M \qquad\qquad \forall i,l : 0 \leq i < l \leq I. \qquad\qquad (4.88)$$

Similarly, the frequency of the low-pass filtered mixed signal during the measurement downsweep, given by:

$$f_{dn,i}^M < f_{dn,l}^M \qquad\qquad \forall i,l : 0 \leq i < l \leq I, \qquad\qquad (4.89)$$

increases as the index i increases.

Assuming that all peaks in the power spectral densities are well separated from each other, the peaks corresponding to the LOS path between the transponder and the base station can be selected. The peak with the highest frequency $f_{up,0}^M$ is used when evaluating the power spectral density of the low-pass filtered mixed signal during the measurement upsweep. The peak with the lowest frequency $f_{dn,0}^M$ is chosen for the measurement downsweep. If $f_{up,0}^M$ and $f_{dn,0}^M$ are used to calculate the distance between the radar stations there is no systematic error and the measurement results are as accurate as in the case of a single LOS connection between the units. Hence, multipath distortions can be compensated for in the base station.

Two power spectra recorded during the measurement downsweep are shown in figure 4.33. In figure 4.33(a) the peak corresponding to the LOS path is the peak with the strongest level. However, in figure 4.33(b) the global maximum is set by the level of an NLOS signal component. Again, there is no guarantee that the LOS signal is the strongest of the arriving signals [63]. Hence, the selection of the peaks must not be based on the signal level alone. If the peak with the lowest frequency is detected in the power spectral density for the measurement downsweep, then the distance from the base station to the transponder is always measured correctly.

The novel method for the mitigation of multipath distortions presented here has been tested in an industrial environment. The measurement results depicted in figure 4.34 are obtained during a measurement campaign in a steel mill. The LPR base station is mounted on top of an overhead crane operating below the roof of the building. The transponder is mounted on the wall opposite the crane. Both stations are connected to antennas with a horizontal 3 dB beam width of 65° [37]. The distance of the crane to the wall is monitored continuously by the LPR and a laser ranging system is used to check the measurement results.

For the measurement results shown in figure 4.34(a) the peak selection is based on the level of the peak alone. Here, the highest peak in the power spectral densities is selected deliberately during synchronization and distance measurement, regardless of its frequency. The crane does not move for the first 70 s. In this position the highest peak in the spectrum corresponds to the LOS path and the distance between the crane and the wall is estimated correctly. However, if the crane is in motion, occasionally, the level of one of the NLOS signal components is larger than the level of the LOS path and the wrong peak is selected. Then, the longer distance corresponding to the NLOS path is measured by the LPR and the true distance is estimated with an error of up to 6 m.

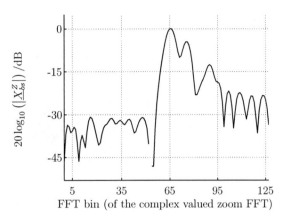

(a) Highest level at the peak corresponding to the LOS path.

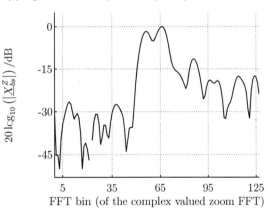

(b) Highest level at a peak corresponding to an NLOS path.

Figure 4.33: Selection of the peak corresponding to the LOS path (measured data, base station): (a) The peak corresponding to the LOS path is the highest peak. (b) However, the global maximum can be determined by the level of a NLOS signal component as well. In any case, the peak corresponding to the LOS path has to be used for the measurements. Here, during the measurement downsweep, this is the peak with the lowest frequency.

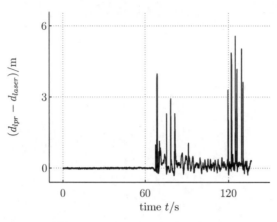

(a) Peak with the highest level is selected.

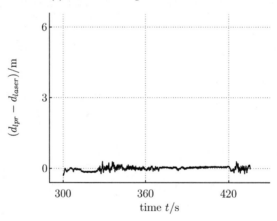

(b) Peak corresponding to the LOS path is selected.

Figure 4.34: Mitigation of multipath distortions (measured data, base station): The distance of a crane to a wall is measured continuously and the measurement error of the LPR with respect to a laser ranging system is monitored over time. (a) At the beginning the peak with the highest level is chosen deliberately. During the first 70 s the crane does not move and the highest peak matches the peak corresponding to the LOS path. After $t \approx 70$ s the crane moves continuously. Occasionally, the level of a peak corresponding to a NLOS component is the highest level. Then, measurement errors of up to 6 m occur. (b) After $t = 300$ s the peak corresponding to the LOS path is selected based on its frequency rather than on its level alone. Consequently, no large measurement errors occur.

After $t = 300\,$s the novel peak selection criterion is used. As derived above the peaks are now selected based on their position in the power spectral densities with respect to the other peaks. In each power spectral density the peak corresponding to the LOS path is chosen based on its frequency even if its level is lower than the level of the NLOS components. The results are shown in figure 4.34(b). No large measurement errors occur although the crane is in motion almost all the time. However, measurement errors of up to 30 cm occur since the Doppler frequency shift has not been compensated for during this particular measurement.

4.9.2 Axial Resolution of the Radar System

In the previous section it has been assumed that the peaks in the power spectral densities corresponding to the LOS and NLOS components of the received signal are well separated from each other. However, if the difference in length between the LOS path and the shortest NLOS path is too short the respective peaks in the power spectral densities interfere with each other.

Due to the limited duration and the limited bandwidth of the signals the shape of a single peak in the power spectral density can be approximated by the probability density function of a Gaussian random variable [31]. Vossiek [99, 100] and Gulden [31] have shown that two Gaussian peaks in the power spectral density can be resolved if they are separated by at least:

$$\Delta f_p \approx \frac{1 + \hat{\xi}_n}{1.2} \frac{F_f}{T}, \tag{4.90}$$

where $\hat{\xi}_n$ is the normalized noise level with respect to the amplitude, F_f is a constant factor depending on the window function used with the digitized data and T is the duration of the sampled signals. Therefore, if the sweep and sampling parameters of the current system setup are used, two peaks can be resolved if they are separated by at least:

$$\Delta f_{p,min} = 1.46\,\text{kHz.} \tag{4.91}$$

This corresponds to approximately 6 bins of the zoom FFT introduced in section 4.2.2.

The minimum spacing $\Delta f_{p,min}$ of the peaks is obtained from (4.90) if the normalized noise level is set to an idealized minimum value of $\hat{\xi}_n = 0$. The sweep duration is $T = 0.9845\,$ms in practice. Furthermore, $F_f = 1.73$ is assumed since a Blackman window is applied to the digitized data. Values of F_f for the Hamming window and the rectangular window have already been given in table 4.1.

The best axial resolution could be achieved if a rectangular window was used. However, in section 4.2.4 and section 4.7.2 it already has been shown that the accuracy of the frequency estimation is improved if a Blackman window is used. This especially holds if the sweep rate of the transponder does not match the sweep rate of the base station.

A minimum difference in length between the LOS and the shortest NLOS path can be derived from the minimum spacing $\Delta f_{p,min}$ of the peaks in the power spectral densities. It depends on the bandwidth B of the FMCW radar sweeps and is given by:

$$\Delta d_{nlos-los,min} \;=\; \Delta f_{p,min} \frac{c_{ph}T}{B} \tag{4.92}$$

$$=\; c_{ph} \frac{1+\hat{\xi}_n}{1.2} \frac{F_f}{B}. \tag{4.93}$$

As the bandwidth increases shorter multipath components can be resolved. For the hardware setup presented in chapter 3 the bandwidth is $B = 132\,\text{MHz}$. The phase velocity of the radar signals is $c_{ph} \approx 3 \cdot 10^8\,\text{m/s}$. Hence, multipath components can be resolved if the NLOS path is longer than the LOS path by at least:

$$\Delta d_{nlos-los,min} \approx 3.3\,\text{m}. \tag{4.94}$$

This is also referred to as the axial resolution of the radar system.

If the difference in distance between the LOS and the NLOS path is shorter than $\Delta d_{nlos-los,min}$, then the peaks in the power spectral densities cannot be resolved. The estimation of the frequency of the low-pass filtered mixed signal will then be subjected to an error. Even if the peaks are resolvable they can interfere with each other. The interference of the LOS and the NLOS signal components results in an estimation error of the frequency corresponding to the LOS signal.

The estimation error depends on the difference in frequency and the offset in phase of the LOS and the NLOS signal components. However, the offset in phase can change rapidly if the radar stations move relatively to each other. Thus, the estimation error of the frequency of the low-pass filtered mixed, which is caused by multipath distortions during the synchronization upsweep, can differ randomly from the estimation error during the synchronization downsweep and the measurement sweeps. Therefore, the frequency estimation error, which is investigated by simulations in the following, is not converted to a corresponding estimation error of the offsets in time and in frequency of the radar units.

For the simulations an LOS and a single NLOS propagation path are assumed. Consequently, the low-pass filtered mixed signal is comprised of two sinusoidal signals which correspond to the LOS and the NLOS propagation path. The frequency of the sinusoidal signal corresponding to the LOS path is $f_{dn,0}^M$, its phase is $\varphi_{dn,0}^M$. Similarly, $f_{dn,1}^M$ and $\varphi_{dn,1}^M$ are the frequency and the phase of the component of the low-pass filtered mixed signal corresponding to the NLOS path. The amplitude of both sinusoidal signals is set to 1 and it is assumed that the LOS and NLOS signal components are of equal magnitude.

The difference in frequency of the signals is changed from $0\,\text{kHz}$ to $3\,\text{kHz}$ in steps of $5\,\text{Hz}$ and the phase offset is changed from 0 to 2π in steps of $\pi/360$, i. e. in steps of $0.5\,°$. For each set of simulation parameters the algorithm for frequency estimation described in section 4.2 is applied to the sum of the two sinusoidal signals. The simulation results are shown in figure 4.35. Similar results are reported by Götz [30].

Figure 4.35(a) depicts the estimation error of the frequency component of the low-pass filtered mixed signal that corresponds to the LOS path. The error is within $\pm 1\,\text{kHz}$. At

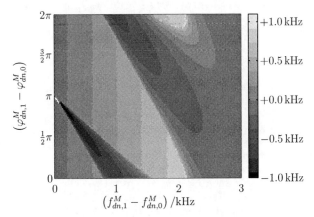

(a) Estimation error of the frequency of the LOS component.

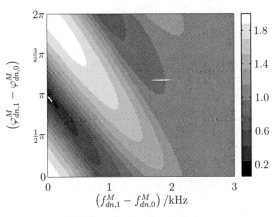

(b) Normalized peak magnitude.

Figure 4.35: Interference of short multipath components (simulated data): If the frequency components $f_{dn,0}^M$ and $f_{dn,1}^M$ of the low-pass filtered mixed signal corresponding to the LOS and the NLOS propagation path, respectively, are close to each other the respective peaks in the power spectral densities interfere. (a) $f_{dn,0}^M$ is then estimated with an error of up to $\pm 1\,\text{kHz}$. (b) The magnitude of the peak in the power spectral density can be increased by a factor of two or reduced to almost zero. The errors depend on the offset in phase of the sinusoidal signals and the difference of their frequencies.

a sweep bandwidth of 132 MHz and a sweep duration of 0.9845 ms this corresponds to a distance measurement error of up to ±1.12 m, if the offsets in time and in frequency have been estimated correctly during synchronization. Otherwise, synchronization errors can increase the distance measurement error even further. In figure 4.35(a) the frequency estimation error is negligible if the difference in frequency of the sinusoidal signals is larger than 2.6 kHz. This corresponds to a difference in distance between the LOS and the NLOS signal of 5.8 m.

The amplitude of the peaks detected in the power spectral densities is shown in figure 4.35(b). It is normalized to the amplitude of the original sinusoidal signal corresponding to the LOS path. Depending on the offset in phase and the difference in frequency of the sinusoidal signals the peak magnitude can be increased to twice the original size or be reduced to a value close to zero. If the magnitude is reduced the SNR of the low-pass filtered mixed signal decreases as well. In the worst case scenario the peak in the power spectral density, which corresponds to the LOS signal component, is eliminated completely due to interference with the NLOS signal. In figure 4.35(b) the error due to the interference of the sinusoidal signals is negligible if the difference in frequency of the sinusoidal signals is larger than 2.2 kHz. This corresponds to a difference in distance between the LOS and the NLOS signal of 4.9 m.

The estimation error of the frequency of the low-pass filtered mixed signal depends on both the difference in frequency and the offset in phase of the LOS and NLOS signal component. For practical application the dependence on the offset in phase is more relevant. If the difference in distance between two propagation paths changes within a single wavelength of 5.2 cm the offset in phase between the respective signals changes within the entire interval $[0, 2\pi)$. However, the difference in frequency changes by less than 25 Hz. Hence, the offset in phase changes much more rapidly than the difference of the frequencies of the signal components if the radar stations move in a multipath environment. Therefore, changes in the offset in phase are the dominant source of error in multipath environments.

The simulation results depicted in figure 4.35 imply that the synchronization and measurement error caused by interference of the multipath components can be neglected completely if the difference in distance between the LOS and NLOS path is larger than 5.8 m. This threshold is inversely proportional to the bandwidth B of the radar signals. Therefore, shorter multipath components can be resolved if the bandwidth of the synchronization and measurement sweeps is increased. However, the bandwidth of the system presented in this thesis is limited to 132 MHz, since a total bandwidth of only 150 MHz is available in the ISM band at 5.8 GHz and some bandwidth has to be reserved for FSK communication.

The errors caused by short multipath components can be mitigated to some degree by advanced signal processing. The algorithm presented in this thesis uses the position of the peaks in the power spectral densities to estimate the frequency of the low-pass filtered mixed signal. However, Götz has shown that the synchronization and measurement results of the measurement system become less sensitive to multipath distortions if the frequency of the low-pass filtered mixed signal is estimated from the points of inflection of the power spectral densities [30]. After the point of inflection is detected the width of an ideal, i. e. undistorted, peak is used to calculate the ideal position of the maximum. Thereby, the effect of the widening of the peaks due to interference of multipath components and the subsequent shift of the position of the maximum is counteracted.

However, the application of the algorithm derived by Götz to the system presented in this thesis is somewhat limited, since the width of the peaks does not only increase due to interference of multipath components. It has been shown in section 4.7 that the width of the peaks increases as well if the sweep rates of the base station and the transponder do not match. Therefore, the ideal peak width for LOS signal propagation conditions depends on the mismatch of the sweep rates of the radar stations, and thus on the deviation of their respective clock frequencies.

4.10 Summary

In this chapter possible sources of error have been identified. The most significant sources of error are summarized in the following.

In dense multipath environments the performance of the measurement system is limited by multipath propagation. If multiple copies of the transmitted signal arrive at the receiver the respective components of the low-pass filtered mixed signal interfere with each other. The peaks in the power spectral densities corresponding to the LOS component are shifted from their original position if the difference in length of the LOS and the NLOS path is below 5.8 m. The exact measurement error depends on the offset in phase, the difference in frequency, and the amplitude of the LOS and NLOS signal components. Simulation results imply measurement errors of typically 0.2 m to 1 m. Multipath resolution could be improved by increasing the bandwidth of the system. At a higher bandwidth, shorter multipath components can be resolved and do not affect the peaks corresponding to the LOS path anymore. Furthermore, the standard deviation of the distance measurements is reduced slightly if the bandwidth is increased.

If only LOS propagation is assumed the accuracy of the system is limited clearly by the hardware components and not by any theoretical limits. For instance, the sweep rate of the transponder cannot be adjusted to the sweep rate of the base station due to the resolution of the DDS. The sweep rates of the stations are linked directly to their respective clock frequencies. The stability of the clock oscillators is ±25 ppm. Therefore, the deviation of the clock frequencies of the base station and the transponder is up to 50 ppm. As the absolute deviation of the sweep rates increases the peaks in the power spectral densities widen and the estimation of the frequency of the low-pass filtered mixed signal becomes more inaccurate. The standard deviation of the distance measurements increases from 6.3 mm if the sweep rates match to 30 mm at a deviation of the clock frequencies of ±50 ppm. The mismatch of the sweep rates can only be overcome by redesigning the hardware. The sweep rate is adjustable if either the resolution of the DDS is increased or a VCXO is used to clock the system. Alternatively, oscillators with a frequency stability better than ±25 ppm can be used to reduce the maximum error due to the mismatch of the sweep rates.

If the sweep rates of both stations match the system performance is limited by the SNR of the low-pass filtered mixed signal. For short distances the SNR of the low-pass filtered mixed signal is given by the phase noise of the PLL used to generate the FMCW radar signals, the standard deviation is then 6.3 mm. For longer distances the standard deviation increases as the level of the received signal, and hence the SNR of the low-pass filtered mixed signal decreases. The maximum range of the system and the range where a low standard deviation is achieved depend on the gain of the antennas used with the system. Estimates are given in table 5.3 in the next chapter.

Additional measurement errors occur if the sampling frequency, i.e. the clock frequency, of the base station or the phase velocity of the radar signals is not known exactly. Both errors are linear in the deviation of the clock frequency or phase velocity from their respective nominal values. The clock frequency can deviate from its nominal value by up to ±25 ppm for the current hardware setup. Furthermore, the phase velocity of the radar signals changes by up to ±150 ppm over environmental conditions for typical application scenarios. However, measurement errors due to an inaccurate sampling frequency or phase velocity of the radar signals can be overcome by calibrating the system at a known distance. Since the errors are linear in the measured distance a large distance should be used for calibration.

Finally, the level of the low-pass filtered signal must be monitored. If the level of the signal is too high the signal is clipped, and harmonics of the fundamental frequency of the low-pass filtered mixed signal appear in the power spectral densities. This causes measurement errors if the spectrum of the low-pass filtered mixed signal is divided into multiple measurement channels to simultaneously measure the distance from the base station to multiple transponders. However, a step attenuator has been included in the current hardware design. Therefore, clipping of the low-pass filtered mixed signal can be avoided since the transmitting station can easily adjust the level of the transmitted signal.

If the sources of error identified here are monitored closely the measurement system presented in this thesis allows for highly accurate distance and velocity measurements. The results of an extensive measurement campaign are presented in the next chapter.

Chapter 5 — Experimental Evaluation of the Measurement System

The local positioning radar presented in the previous chapters is tested in various environments. Four different test setups are used to verify the excellent performance of the distance measurement system at hand.

In the initial scenario in section 5.1 the base station and the transponder are connected by a delay line. Therefore, any influences from the environment like multipath propagation can be neglected. In this setup the radar signals can be attenuated easily and the effect of the SNR on the accuracy of the measurement results can be shown. Furthermore, the dependence of the measurement results on the deviation of the clock frequencies can be investigated without any distortions due to the environment.

Subsequently, the LPR is tested at an airfield. The results are presented in section 5.2. Here, line-of-sight propagation conditions can be assumed as well for short distances. For long distances ground reflections are to be expected. However, the influence of the environment is still at a minimum. At the airfield the system is tested over large distances of more than 500 m. The measurement results are compared to the results of an industrial laser ranging system to verify their accuracy.

Finally, distance measurements obtained in an indoor environment are presented in section 5.3 and section 5.4. In section 5.3 the laser ranging system is used to verify the measurement results again. An even more precise reference is provided by a linear unit in section 5.4. Here, the effect of clutter noise from the environment can be shown.

In the following, the measurement results obtained in the four test scenarios are presented. They are compared to the simulation results in chapter 4 where applicable. The system parameters given in table 3.1, e.g. a sweep bandwidth of 132 MHz and a sweep duration of 0.9845 ms, are used for all measurements.

5.1 Delay Line Measurements

In the first test setup the base station and the transponder are connected by a delay line. Since the radar stations are transmitting signals in the ISM band at 5.8 GHz, any delay line certified for frequencies up to 6 GHz could be used for the tests. A delay line of type LL2773-AF [20] has been chosen arbitrarily. The mechanical length of the delay line is approximately 100 m. However, the distance which is measured through the delay line is the electrical length of the delay line. It is given by the mechanical length and the relative permittivity of the delay line (4.75) [20]. The electrical length of the delay line is approximately 120 m. The measured distance is increased by a few meters due to the cables used to connect the radar units to the delay line. Therefore,

delay line

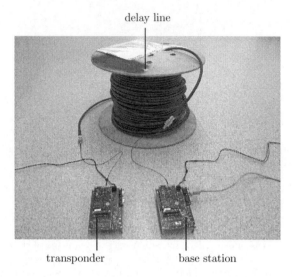

transponder   base station

Figure 5.1: Delay line, setup: The base station and the transponder are connected by a delay line. Hence, the electrical length of the delay line is measured.

distance measurements with a mean value between 125 m and 130 m are expected in the measurement setup depicted in figure 5.1.

Since the radar stations are connected by a delay line the propagation channel is well-known. Multipath distortions due to the environment can be neglected completely. Clearly, there is no Doppler frequency shift since the radar stations do not move relatively to each other. Furthermore, the attenuation of the radar signals is known exactly from the electrical specifications of the delay line [20]. It can be increased further by an additional attenuator to investigate the dependence of the measurement results on the signal level.

5.1.1 Distribution of the Distance Measurements

To evaluate the accuracy of the LPR system the distribution of the measurement error must be known. Therefore, the probability density function of the distance measurements is obtained using 1.75 million samples of the electrical length of the delay line. Figure 5.2 depicts the distribution of the measured distance (black line) and the probability density function of a Gaussian random variable with the same mean and standard deviation (gray line).

Both graphs match each other well which implies a Gaussian distribution of the distance measurements. The standard deviation of the samples is 6.88 mm. It is slightly larger than the minimum standard deviation of 6.3 mm obtained in the simulations in

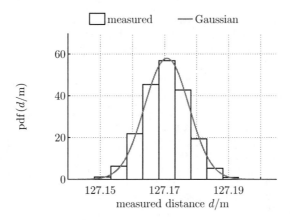

Figure 5.2: Delay line, distance measurements: 1.75 million measurements of the length of the delay line are taken. The probability density function (pdf) of the distance measurements approximates a Gaussian distribution. The standard deviation is only 6.88 mm.

the previous chapter since the clock frequencies of the base station and the transponder, and therefore the sweep rates of both stations, do not match.

A χ^2 goodness-of-fit test [12, 13] is performed to test for a Gaussian distribution of the recorded samples. The data are grouped into $I = 13$ bins. For each bin the observed counts $n_{o,i}$ are calculated subsequently from the measured distances. The expected counts $n_{e,i}$ for those bins are obtained from a Gaussian distribution with mean 127.1704 m and standard deviation 6.88 mm. The test results are shown in table 5.1.

The χ^2 test statistic, given by:

$$\chi^2_s = \sum_{i=1}^{13} \frac{(n_{o,i} - n_{e,i})^2}{n_{e,i}} = 13.6315, \qquad (5.1)$$

is calculated subsequently. It has to be compared to the quantile of the χ^2 distribution with 10 degrees of freedom corresponding to the 5 % significance level [12]. The mean and standard deviation of the Gaussian distribution, which the measurement results are tested for, are obtained from the measured samples. Therefore, the number of degrees of freedom of the χ^2 distribution is given by the number of bins I the measurement results have been grouped into reduced by three [12].

The quantile of the χ^2 distribution with 10 degrees of freedom corresponding to the 5 % significance level is 18.307 [12]. Clearly, the test statistic is smaller than the respective quantile of the χ^2 distribution, i.e.:

$$\chi^2_s < 18.307. \qquad (5.2)$$

Therefore, the null hypothesis, i.e. the distance measurements are random samples from a Gaussian distribution, cannot be rejected at the 5 % significance level [12].

Table 5.1: χ^2 goodness-of-fit test of the distance measurements: The samples of the measured length of the delay are tested for a Gaussian distribution with mean 127.1704 m and standard deviation 6.88 mm. This null hypothesis cannot be rejected at the 5 % significance level.

index i	lower limit of bin i in m	observed counts $n_{o,i}$	expected counts $n_{e,i}$	$\dfrac{(n_{o,i} - n_{e,i})^2}{n_{e,i}}$
1	$-\infty$	80	67	2.5224
2	127.1432	1050	1028	0.4708
3	127.1482	9874	9771	1.0858
4	127.1532	55369	55796	3.2678
5	127.1582	191098	191746	2.1899
6	127.1632	397094	397092	0.0000
7	127.1682	496926	495977	1.8158
8	127.1732	374022	373724	0.2376
9	127.1782	169543	169830	0.4850
10	127.1832	46464	46501	0.0294
11	127.1882	7677	7661	0.0334
12	127.1932	748	758	0.1319
13	127.1982	55	47	1.3617
	∞			

$$\sum_{i=1}^{13} \frac{(n_{o,i} - n_{e,i})^2}{n_{e,i}} = 13.6315$$

Consequently, a Gaussian distribution of the distance measurement error of the LPR can be assumed for line-of-sight environments.

5.1.2 Distribution of the Velocity Measurements

Similarly, the probability density function of the velocity measurements is obtained from 850000 samples of the measured velocity. Figure 5.3 depicts the distribution of the measured velocity (black line) and the probability density function of a Gaussian random variable with the same mean and standard deviation (gray line). The standard deviation of the samples is 0.175 m/s. Both graphs match each other well which implies a Gaussian distribution of the velocity measurements.

A χ^2 goodness-of-fit test [12,13] is performed to test for a Gaussian distribution of the recorded samples. The data are grouped into $I = 11$ bins. For each bin the observed counts $n_{o,i}$ are calculated subsequently from the measured velocities. The expected counts $n_{e,i}$ for those bins are obtained from a Gaussian distribution with mean 0.05 m/s and standard deviation 0.1751 m/s. The test results are shown in table 5.2.

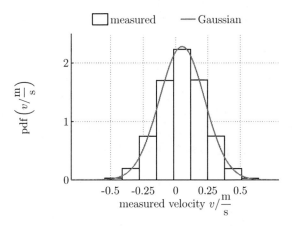

Figure 5.3: Delay line, velocity measurements: 850000 measurements of the velocity are taken. The probability density function (pdf) of the velocity measurements approximates a Gaussian distribution. The standard deviation is only 0.175 m/s.

Table 5.2: χ^2 goodness-of-fit test of the velocity measurements: The samples of the measured velocity are tested for a Gaussian distribution with mean 0.05 m/s and standard deviation 0.1751 m/s. This null hypothesis cannot be rejected at the 5 % significance level.

index i	lower limit of bin i in m/s	observed counts $n_{o,i}$	expected counts $n_{e,i}$	$\dfrac{(n_{o,i} - n_{e,i})^2}{n_{e,i}}$
1	$-\infty$	23	17	2.1176
2	-0.67	612	571	2.9440
3	-0.51	8997	8910	0.8495
4	-0.35	62800	62959	0.4015
5	-0.19	202589	202840	0.3106
6	-0.03	299360	299407	0.0074
7	0.13	203429	202840	1.7103
8	0.29	62686	62959	1.1838
9	0.45	8908	8910	0.0004
10	0.61	571	571	0.0000
11	0.77	25	17	3.7647
	∞			

$$\sum_{i=1}^{11} \frac{(n_{o,i} - n_{e,i})^2}{n_{e,i}} = 13.2898$$

The χ^2 test statistic, given by:

$$\chi_s^2 = \sum_{i=1}^{11} \frac{(n_{o,i} - n_{e,i})^2}{n_{e,i}} = 13.2898, \tag{5.3}$$

is calculated subsequently. It is compared to the quantile of the χ^2 distribution with 8 degrees of freedom corresponding to the 5 % significance level. Since the mean and standard deviation of the Gaussian distribution have been estimated from the measured data, the number of degrees of freedom of the χ^2 distribution is given by the number of bins I the data have been grouped into reduced by three [12].

The quantile of the χ^2 distribution with 8 degrees of freedom corresponding to the 5 % significance level is 15.507 [12]. Clearly, the test statistic is smaller than the respective quantile of the χ^2 distribution, i. e.:

$$\chi_s^2 < 15.507. \tag{5.4}$$

Therefore, the null hypothesis, i. e. the velocity measurements are random samples from a Gaussian distribution, cannot be rejected at the 5 % significance level [12]. Consequently, a Gaussian distribution of the velocity measurement error of the LPR can be assumed for line-of-sight environments.

Note, that the velocity measurements are expected to be zero mean if the stations are connected by a delay line. However, an average velocity of approximately 0.05 m/s is measured here. The offset is caused most likely by the resolution of the DDS. It is observed in the measurements in section 5.1.4 as well, if the deviation of the clock frequencies of the transponder and the base station is larger than 10 ppm.

In the previous sections it has been shown that the distance and velocity measurements of the LPR are samples from a Gaussian distribution. However, the standard deviation of the measurement results depends on the level of the received signal. This is shown next.

5.1.3 Dependence on the Level of the Received Signal

The simulation results presented in section 4.5.5 have shown that the accuracy of the measurement results depends on the effective SNR of the low-pass filtered mixed signal. The effective SNR mainly depends on the phase noise of the PLL in the signal generator and the attenuation of the signals during transmission.

A delay line is used to measure the dependence of the standard deviation of the measurement results on the level of the received signal. The measurement setup depicted in figure 5.1 is modified slightly. The base station and the transponder are connected by a shorter delay line and a step attenuator [58] is added to the transmission path. The step attenuator is shown in figure 5.4. Its attenuation a_{sa} can be changed from 0 dB to 69 dB.

The total attenuation a_{path} of the radar signals during transmission is comprised of the attenuation a_{sa} of the step attenuator and the attenuation a_{dl} of the delay line. It is given by:

$$a_{path} = a_{dl} + a_{sa}. \tag{5.5}$$

Figure 5.4: Step attenuator, setup: The base station and the transponder are connected by a short delay line. A step attenuator is connected to the delay line which additionally attenuates the signal by up to 69 dB.

Since the attenuation of the delay line is 25 dB, the total attenuation of the transmission line ranges from 25 dB to 94 dB. For the measurement results depicted in figure 5.5 and figure 5.6 the total attenuation is increased in steps of 2 dB starting at 25 dB. For each attenuation 900 samples of the length of the delay line are acquired.

Mean and Standard Deviation of the Distance Measurements

The mean value of the measured distances is shown in figure 5.5. It does not change significantly with respect to the attenuation. The mean values are within ±10 mm from their average. Minor changes can be caused by the step attenuator itself since the attenuation is adjusted mechanically. Naturally, the measured distance changes when the electrical length of the step attenuator varies if the attenuation is changed.

Figure 5.6 depicts the standard deviation of the acquired distances. As long as the total attenuation of the transmission line is below 67 dB the standard deviation is 6.9 mm. It matches the standard deviation of the distance measurements in section 5.1.1. When the attenuation of the radar signals is increased beyond 67 dB the standard deviation of the distance measurements increases. At the maximum allowable attenuation of 93 dB the standard deviation is 40 mm. If the total attenuation of the transmission line is increased to 94 dB the level of the peaks in the power spectral densities is too low to be detected reliably.

The measurement results (black line) match the simulation results (gray line) presented in figure 4.19. As long as the total attenuation is below 67 dB the effective SNR γ_e near the peak in the power spectral density is determined by the phase noise of the PLL of the current hardware setup (4.37). Given the FFT parameters discussed in section 4.2 the peaks in the power spectral densities are approximately 50 dB above the noise floor caused by the phase noise of the PLL.

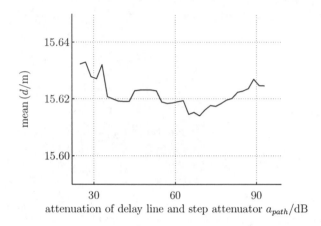

Figure 5.5: Step attenuator, mean of the distance measurements: The attenuation of the delay line is $a_{dl} = 25$ dB. The attenuation of the step attenuator is increased from 0 dB to 68 dB in steps of 2 dB. Therefore, the total attenuation of the transmission line increases from 25 dB to 93 dB. Approximately 900 measurements of the distance are taken for each attenuation. The mean value of the samples does not change significantly.

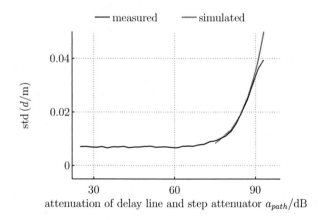

Figure 5.6: Step attenuator, standard deviation of the distance measurements: The standard deviation of the distances measured by the LPR is approximately 6.9 mm for an attenuation $a_{path} \leq 67$ dB. Here, the phase noise of the PLL is the dominant noise source. As the attenuation increases beyond 67 dB the standard deviation increases as well. The maximum allowable attenuation is 93 dB. Then, the standard deviation is 40 mm.

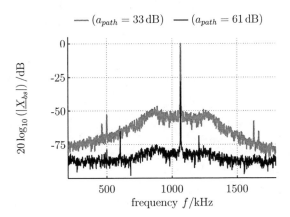

Figure 5.7: Step attenuator, effective SNR (measured data, base station): For low attenuations (gray line), the effective SNR γ_e is given by the phase noise of the PLL of the current hardware setup. As the attenuation increases both the peak power and the noise floor caused by the PLL decrease (black line). Therefore, the effective SNR remains constant. If the attenuation increases beyond 67 dB only the peak power decreases while the noise level remains constant. Consequently, the effective SNR is reduced and the standard deviation of the measurements increases.

Exemplary power spectra are shown in figure 5.7. Clearly, the phase noise of the PLL dominates the noise level near the peak for low attenuations (gray line). As the attenuation is increased both the peak power and the noise floor caused by the PLL phase noise decrease (black line). Hence, the effective SNR and therefore the standard deviation of the distance measurements remain constant. However, if the total attenuation is increased beyond 67 dB the effective SNR is determined by the thermal noise of the receiver. Then, the noise level remains constant and only the level of the peak decreases. Consequently, the effective SNR is reduced and the standard deviation of the distance measurements increases.

Maximum Transmission Range of the System

The maximum range of the measurement system with respect to the signal level can be estimated from the measurements presented above. The total attenuation of the delay line and the step attenuator correspond to the path loss of the wireless channel between the base station and the transponder. The allowable distance between the radar stations is calculated subsequently from the free-space path loss (4.35). At a frequency of 5.8 GHz a path loss of 67 dB corresponds to a range of 9 m where a low standard deviation of the distance measurement results is achieved. Similarly, a path loss of 93 dB equals a distance of 184 m. This is the maximum range of the measurement system if antennas without any directional gain are used.

Table 5.3: Step attenuator, estimated maximum range of the system: The maximum range of the LPR depends on the antennas used with the system. For high gain antennas with a gain of 23 dBi the range is up to 4 km. A low standard deviation of the measurements of 6.9 mm can be achieved if the distance is below 200 m.

	delay line	antenna (6 dBi)	antenna (12 dBi)	antenna (23 dBi)
tx gain / dBi		5	5	5
rx gain / dBi		5	11	22
low std				
path loss / dB	67	77	83	94
distance / m	9	29	58	206
maximum range				
path loss / dB	93	103	109	120
distance / m	184	581	1160	4116

In practice, the measurement system is used with a variety of antennas [94]. Then, the range of the measurement system is increased due to the directional gain of the antennas. Currently, the antennas most often used with the system are omni-directional antennas (6 dBi gain, [48]) with a vertical 3 dB beam width of 8.5°, planar antennas (12 dBi gain, [37]) with a horizontal 3 dB beam width of 65° and a vertical 3 dB beam width of 35°, or planar antennas (23 dBi gain, [38]) with a horizontal and vertical 3 dB beam width of 9°.

The directional gain of the antennas used with the system has to be included in the calculations of the maximum measurement range. The maximum transmit power within the ISM band at 5.8 GHz is limited to 14 dBm EIRP by legal requirements [68]. Since the maximum output power of an LPR station is 9 dBm a gain of 5 dBi can be used on the transmitting station. If the directional gain of the antennas exceeds 5 dBi the step attenuator included in the current hardware design has to be used to reduce the transmit power to the allowable level. However, the entire directional gain of the antennas can be used on the receiving station. For the calculations shown in table 5.3 the gain of the antennas of the receiving station has been reduced by 1 dBi to account for the attenuation of the cables used to connect the radar stations to the antennas.

The allowable path loss of the wireless channel between the base station and the transponder increases by the gain due to the antennas used with the transmitting and the receiving station. Consequently, the transmission range of the system increases as well. If a gain of 5 dBi is assumed for the transmitting station then the level of the transmitted signal matches the maximum allowable transmit power in the ISM band of 14 dBm EIRP [68]. Then, the maximum transmission range of the system is found to be 581 m, 1160 m, or 4116 m if directional gains of 5 dBi, 11 dBi, or 22 dBi are assumed for the antenna of the receiving station, respectively. For distances up to 29 m, 58 m, or 206 m, respectively, a low standard deviation of 6.9 mm can be achieved. Here, the system performance is limited by the phase noise of the PLL of the signal generator.

In section 5.1.4 it will be shown that the potential mismatch of the sweep rates of the radar stations then limits the accuracy of the measurement results.

If high gain antennas are used with the system the transmission range of the system exceeds 4 km if only the level of the signals is considered. However, in section 2.3.3 it has already been shown that the measurement range of the system is limited by the frequency the low-pass filtered mixed signal is digitized with in the base station and by the size of the FFT used to evaluate the digitized signal. Currently, the measurement range is limited to approximately 2.23 km (4.27). It could be increased to more than 4 km if the distance and velocity measurements were evaluated according to the extended algorithm described in section 2.7.

5.1.4 Dependence on the Clock Frequency

The standard deviation of the measurement results also depends on the mismatch of the sweep rates of the base station and the transponder and, consequently, on the deviation of their respective clock frequencies. This has already been shown by the simulations presented in section 4.7.4. The simulation results are now verified by measurements.

For the measurements the hardware configuration of the transponder is modified. The crystal oscillator is removed from the printed circuit board and the transponder is connected to a signal generator. The clock frequency of the transponder can now be adjusted by setting the output frequency of the signal generator. The clock frequency of the base station is still given by the frequency of the on-board crystal oscillator.

The performance of the measurement system strongly depends on the phase noise of the signal generator used to clock the transponder. If the phase noise of the generator is too high the effective SNR of the low-pass filtered mixed signal in the base station and the transponder decreases and the standard deviation of the measurement results increases. Consequently, the phase noise of the 5.8 GHz signal of the modified transponder must not exceed the phase noise of the signal of an unmodified station. Otherwise, the measurement results cannot be adopted directly to the standard hardware setup.

Figure 5.8 depicts the phase noise of a 5.8 GHz continuous wave signal of the conventional LPR (black line) and the phase noise of the signal of the modified transponder (gray line). For the measurements the transponder is clocked by the Anritsu signal generator MG3691B [6]. The phase noise of the generator clocked transponder then matches the phase noise of an unmodified station. It is between -80 dBc/Hz and -90 dBc/Hz over the entire bandwidth of the loop filter of the PLL of 200 kHz. It decreases further outside the bandwidth of the loop filter.

Mean and Standard Deviation of the Distance Measurements

For the measurements the base station and the transponder are connected by the delay line depicted in figure 5.1 again. The frequency deviation of the transponder is changed from -50 ppm to 50 ppm with respect to the clock frequency of the base station. This corresponds to the frequency deviation occurring between the stations in practice since the stability of the crystal oscillators is ± 25 ppm [47]. For each frequency deviation 1000 samples of the length of the delay line are acquired, before the deviation is increased by

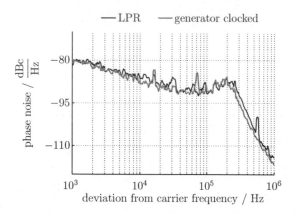

Figure 5.8: Generator clocked LPR, phase noise: The output frequency of the system is set to $f_{rf} = 5.8\,\mathrm{GHz}$. The phase noise of the conventional LPR is between $-80\,\mathrm{dBc/Hz}$ and $-90\,\mathrm{dBc/Hz}$ over the entire bandwidth of the loop filter of 200 kHz (black line). The phase noise of the generator clocked station strongly depends on the phase noise of the signal generator used to provide the clock frequency. If the Anritsu signal generator MG3691B is used to clock the LPR, the phase noise of the signal matches the phase noise of the signal of a conventional LPR (gray line).

1 ppm. The clock frequency of the base station is given by the frequency on the on-board crystal oscillator. It is assumed to remain constant during the entire measurement.

Figure 5.9(a) depicts the mean of the distance measurements acquired for each frequency deviation. It changes by less than 10 mm if the deviation of the clock frequencies is increased from $-25\,\mathrm{ppm}$ to 25 ppm. If the absolute deviation of the clock frequencies exceeds 25 ppm the mean of the distance measurements increases slightly. Due to the mismatch of the sweep rates of the radar stations the estimation of the frequency of the low-pass filtered mixed signal is less accurate. This has been derived in section 4.7 already. However, the mean of the distance measurements is within $\pm 25\,\mathrm{mm}$ for all frequency deviations considered here.

The standard deviation of the distance measurements is shown in figure 5.9(b). It increases as the absolute deviation of the clock frequencies of the base station and the transponder increases. If the clock frequencies of both stations match, the standard deviation is only 6.4 mm. It increases to up to 30 mm as the deviation of the clock frequencies increases to 50 ppm. For deviations within $\pm 25\,\mathrm{ppm}$ the standard deviation is below 10 mm. The measurement results (black line) match the simulation results from figure 4.27 (gray line).

Mean and Standard Deviation of the Velocity Measurements

The dependence of the velocity measurements on the deviation of the clock frequencies of the base station and the transponder is shown with the same measurement setup.

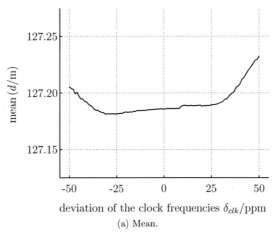

(a) Mean.

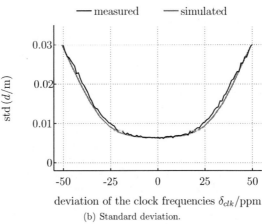

(b) Standard deviation.

Figure 5.9: Dependence of the distance measurements on the clock frequency: The clock frequency of the transponder is changed from $-50\,\mathrm{ppm}$ to $50\,\mathrm{ppm}$ with respect to the clock frequency of the base station in steps of $1\,\mathrm{ppm}$. 1000 measurements of the distance are recorded for each frequency deviation. (a) Their mean values are within a span of $\pm 25\,\mathrm{mm}$ over the entire frequency range. (b) The standard deviation of the distance measurements is $6.4\,\mathrm{mm}$ if the clock rates of the radar stations match. It increases to $30\,\mathrm{mm}$ for the worst case clock frequency deviation of $\pm 50\,\mathrm{ppm}$. The measurement results (black line) match the simulation results from figure 4.27 (gray line).

Again, the deviation of the clock frequencies is increased from -50 ppm to 50 ppm in steps of 1 ppm. For each value 1000 velocity measurements are acquired. Their mean value is shown in figure 5.10(a).

Since the radar stations do not move relatively to each other the velocity measurements are expected to be zero mean. This is true for approximately half of the frequency deviations, e. g. deviations from -40 ppm to 10 ppm. At approximately 10 ppm the mean of the velocity measurements increases to 0.05 m/s. This corresponds to changing the frequency of the measurement sweeps transmitted by the transponder by the equivalent of incrementing the frequency tuning word by 1. If the resolution of the DDS was increased the velocity measurements would most likely be zero mean. Similarly to the mean of the distance measurements the mean of the velocity measurements increases as the deviation of the clock frequencies approaches ± 50 ppm, since the estimation of the frequency of the low-pass filtered mixed signal becomes more inaccurate due to the mismatch of the sweep rates.

The dependence of the standard deviation of the velocity measurements on the deviation of the clock frequencies of the radar stations depicted in figure 5.10(b) is similar to the dependence of the standard deviation of the distance measurements shown in figure 5.9(b). The standard deviation of the velocity measurements is 0.16 m/s if the clock frequency of the base station matches the clock rate of the transponder. It increases to 0.7 m/s as the deviation of the clock frequencies approaches ± 50 ppm.

From figure 5.9 and figure 5.10 it is apparent that the performance of the measurement system is best if the clock frequencies of the base station and the transponder differ by less than 25 ppm. Therefore, oscillators with a stability of ± 12.5 ppm or better should be used to clock the radar units for high precision applications. Alternatively, the signal generator of the radar stations could be modified such that the sweep rate of the transponder can be adjusted to the sweep rate of the base station as discussed in section 4.4.4.

In the delay line measurements presented above the radar distance measurement system at hand has been tested under ideal line-of-sight propagation conditions. The measurement results imply that the distance measurement error can be modeled by a Gaussian random variable, where the standard deviation depends on the attenuation of the radar signals and the mismatch of the sweep rates of the base station and the transponder. It has been shown that the minimum achievable standard deviation of the distance measurements is 6.4 mm which is very close to the minimum standard deviation of 6.3 mm obtained by simulations in chapter 4. In the following, the system is tested in real-world scenarios.

5.2 Measurements at an Airfield

In the previous section a delay line has been used to evaluate the performance of the local positioning radar LPR for ideal line-of-sight propagation conditions. Next, the measurement system is tested on the runway of an airfield. The length of the runway is 2.2 km. Hence, the radar system at hand can be tested over its entire measurement range. Figure 5.11 depicts the measurement setup.

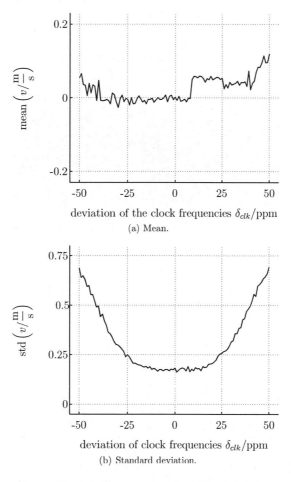

(a) Mean.

(b) Standard deviation.

Figure 5.10: Dependence of the velocity measurements on the clock frequency: The clock frequency of the transponder is changed from -50 ppm to 50 ppm with respect to the clock frequency of the base station in steps of 1 ppm. 1000 measurements of the velocity are recorded for each frequency deviation. (a) The mean of the velocity measurements does not change significantly over the entire range. The sudden increase at approximately 10 ppm corresponds to the change in the frequency of the low-pass filtered mixed signal when the frequency tuning word of the DDS is changed by 1. It is most likely caused by the resolution of the DDS. (b) The standard deviation of the velocity measurements is 0.16 m/s if the clock rates of the radar stations match. It increases to 0.7 m/s for the worst case clock frequency deviation of ± 50 ppm.

tripod with antennas
spotting scope
laser ranging system

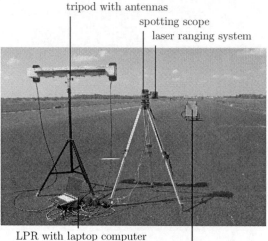

LPR with laptop computer
bicycle with reflector and antennas

Figure 5.11: Airfield, setup: The system is tested on the runway of an airfield. The length of the runway is 2.2 km. The base station is placed at one end of the runway, the base station antenna is mounted on a tripod. The transponder is mounted on a bicycle. The measurement results of the LPR are compared to the distances measured by an industrial laser ranging system. The reflector for the laser is mounted on the bicycle as well. A spotting scope is used to track the bicycle.

The base station is placed at one end of the runway. It is connected to an antenna with a directional gain of 23 dBi [38]. The transmit level of the radar stations is adjusted to 14 dBm to meet the legal requirements for the transmit power [68]. The base station antenna is mounted on a tripod approximately 1.7 m above the ground. The transponder and its antenna are mounted on a bicycle, the antenna of the transponder is approximately 1.2 m above the ground. Line-of-sight only propagation can be assumed for the measurement setup at hand. Occasionally, reflections from the ground will degrade system performance. However, they can be avoided by increasing the clearance of the antennas to the ground.

An industrial laser ranging system [49] is used to check the measurement results of the LPR. Therefore, a reflector is mounted on the bicycle as well. The laser ranging system and a spotting scope are mounted on another tripod close to the base station antenna. The laser ranging system and the spotting scope have to be adjusted properly to allow for a continuous tracking of the bicycle on the runway. Therefore, both the spotting scope and the laser are pointed at the reflector at a distance of approximately 160 m. This is shown in figure 5.12.

After the laser and the spotting scope are adjusted properly the bicycle can be tracked over a distance of up to 550 m. This is the maximum measurement range of the laser

Figure 5.12: Airfield, adjusting the laser ranging system: The laser ranging system has to be
adjusted with high accuracy to track the bicycle. Therefore, the laser ranging
system and the spotting scope are pointed at the reflector at a distance of approximately 160 m. After the laser ranging system and the spotting scope are
adjusted, the bicycle can be tracked over a distance of 550 m.

ranging system. There is no alternative measurement system available to verify the
accuracy of the LPR for distances larger than 550 m.

5.2.1 Distance Measurements

For the measurement results presented next the distance between the base station and
the transponder is measured by the LPR and the laser ranging system. In general the
different height of the antennas of the base station and the transponder as well as the
height of the laser and the reflector have to be taken into account when comparing
the distances measured by both systems. However, for distances larger than 30 m a
mismatch in height of 50 cm causes an error of less than 5 mm. The error is neglected
and the distance measurements obtained by both systems are compared to each other
directly.

Furthermore, there is an offset between the distance measurements of the LPR and
the laser ranging system. The offset is caused by the cables used to connect the base
station and the transponder to their respective antennas. The system is calibrated prior
to the measurements, e.g. the offset is set to zero at a distance of approximately 30 m.

The bicycle is moved away from the base station and the distance measurements
of the LPR and the laser ranging system are monitored continuously. The distance

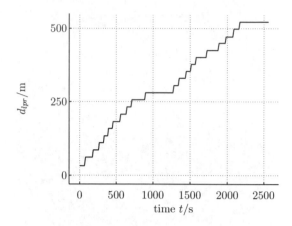

Figure 5.13: Airfield, distance measurements: The bicycle is moved from a distance of 30 m to a distance of 550 m. Occasionally, the bicycle is stopped and at least 1500 measurements of the same distance are taken by the LPR (horizontal lines). The measurement results are compared to values obtained by the laser ranging system.

measurements d_{lpr} acquired by the LPR are shown in figure 5.13. Occasionally, the bicycle is stopped for at least 40 s (horizontal lines). More than 1500 samples of the same distances are recorded by the LPR at each stop before the bicycle is moved again. The distance d_{laser} measured by the laser ranging system is recorded as well. Finally, the laser ranging system stops working at a distance of 550 m and the measurement campaign is stopped.

Figure 5.14 depicts the mean deviation of the distance measurements d_{lpr} and d_{laser} which are acquired at each stop by the LPR and the laser ranging system, respectively. The mean difference is below 30 mm over the entire range of 550 m.

The standard deviation of the distance measurements is shown in figure 5.15. It is below 15 mm at each position. For distances smaller than 200 m the standard deviation is only 6.9 mm. The standard deviation matches the minimum standard deviation obtained in the delay line measurements in section 5.1.1 and section 5.1.3. From table 5.3 the low standard deviation of 6.9 mm is expected for distances up to 206 m if antennas with a directional gain of 23 dBi are used. This is conclusive with the measurement results presented in figure 5.15.

For distances larger than 250 m the standard deviation changes rapidly between 7 mm and 15 mm. This can be caused by either ground echoes or people moving through the propagation path.

Finally, the overall error of the distance measurements of the LPR with respect to the laser ranging system is calculated. The error is defined as the difference of the distances d_{lpr} and d_{laser} estimated by the LPR and by the laser ranging system, respectively. Distances measured over the entire range of 550 m are taken into account. However, distance measurements acquired while the bicycle is moving are ignored here

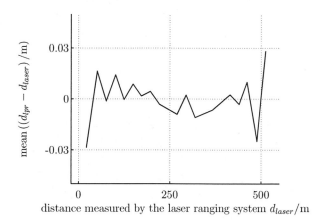

Figure 5.14: Airfield, mean of the distance measurements: The mean difference between the distances measured by the LPR d_{lpr} and by the laser ranging system d_{laser} is below 30 mm over the entire range of 550 m. Note, that the constant offset due to the length of the antenna cables has been subtracted from the distances measured by the LPR as well.

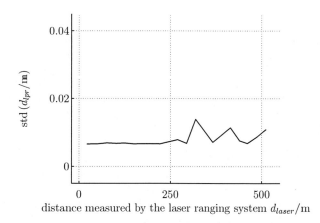

Figure 5.15: Airfield, standard deviation of the distance measurements: The standard deviation of the distances measured by the LPR d_{lpr} is below 15 mm over the entire range. For distances below 200 m the standard deviation is approximately 6.9 mm. The rapid changes in the standard deviation for distances larger than 250 m can be caused by either ground echoes or people moving through the propagation path.

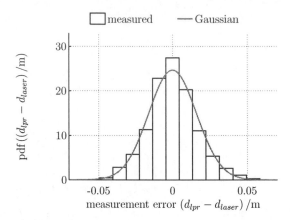

Figure 5.16: Airfield, overall error of the distance measurements: The overall error is defined as the difference between the distances measured by the LPR d_{lpr} and the distances measured by the laser ranging system d_{laser}. The measurement results obtained over the entire range are taken into account. The probability density function (pdf) of the overall error approximates a Gaussian distribution. The standard deviation is only 16 mm.

since the measurements of the LPR and the laser ranging system are not synchronized. The relative motion of the base station and the transponder between the instant the measurement results of the LPR and the laser ranging system are obtained would distort the distribution of the measurement error otherwise. A histogram of the overall error between the measurements of the LPR and the laser ranging system is shown in figure 5.16. The probability density function of the overall error approximates a Gaussian distribution with a standard deviation of only 16 mm.

Both the low standard deviation and the low average error of the distance measurements of the LPR with respect to the measurements of the laser ranging system prove the excellent performance of the algorithms presented in this thesis and the current LPR hardware setup. No major measurement errors occurred over a range of 550 m. In fact, the system has been tested successfully up to a distance of 2.2 km.

Note, that both measurement errors of the laser ranging system and the LPR contribute to the overall error. If the reflector and the antennas at the transponder side are not perpendicular to the line-of-sight to the base station the laser ranging system can measure too large a distance while the antenna of the transponder is closer to the antenna of the base station or vice versa. Therefore, a more precise test setup for short ranges is presented in section 5.4, where a linear unit is used to move the antenna of the base station.

5.2.2 Velocity Measurements

The velocity of the bicycle is measured during another drive along the runway. The bicycle is accelerated to velocities of up to 4.5 m/s and the distance and velocity are monitored continuously. The velocity measurements are depicted in figure 5.17. Note, that the velocity is assumed to be positive if the radar stations approach each other by the definition of the Doppler frequency shift (2.105). Therefore, the negative sign in figure 5.17 indicates that the transponder is moving away from the base station.

In figure 5.17(a) the velocity measurements v_{meas} of the LPR according to (2.130) are represented by the black line. The laser ranging system does not measure the velocity directly. Therefore, the velocity measurements of the laser ranging system are calculated from the change in distance during two consecutive laser measurements Δd_{laser} and the time between those measurements Δt_{laser}. The velocity measurements obtained for the laser ranging system are represented by the gray line. Again, the negative sign is required due to the definition of the Doppler frequency shift (2.105). It indicates an increase in the distance of the stations. The measurement results obtained by the LPR and the laser ranging system match well. Therefore, the relative velocity of the LPR stations can be measured directly from the Doppler frequency shift of the FMCW radar signals.

The velocity of the bicycle can also be calculated from the ratio of the change in distance between two consecutive LPR measurements Δd_{lpr} and the time between those measurements Δt_{lpr}. This is shown in figure 5.17(b). Again, the measurement results of the LPR (black line) match the results calculated from the distance measurements of the laser ranging system (gray line).

5.2.3 Maximum Range of the Radar System

The maximum transmission range of the LPR system cannot be measured directly since it exceeds the length of the runway. However, a good estimate is obtained from the signal level of the low-pass filtered mixed signal at a large distance. Therefore, the base station is placed near one end of the runway. The transponder is set up at the other end of the runway. It is placed on a small hill at a height of approximately 6 m to minimize ground reflections. The distance between the base station and the transponder is approximately 2 km. Both stations are connected to antennas with a directional gain of 23 dBi, the transmit power is limited to 14 dBm which is the maximum allowable transmit power in the ISM band at 5.8 GHz.

Figure 5.18 depicts the power spectral density of the low-pass filtered mixed signal in the base station during the measurement upsweep (f_{up}, gray line) and downsweep (f_{dn}, black line) at a distance of 2 km. The level at f_{dn} is about 5 dB lower than the level at f_{up} since the frequency f_{dn} is close to the cut-off frequency of the low-pass filter in the receiver. This has been discussed in section 4.5.1.

The two peaks in the center of the spectrum are caused by the voltage regulators in the power supply. Clearly, wrong measurement results would be obtained if those peaks were detected during synchronization or distance measurement. Therefore, the level at the frequency of the low-pass filtered mixed signal has to exceed the level of the spurs

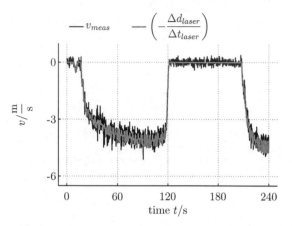

(a) Velocity v_{meas} measured from the Doppler frequency shift.

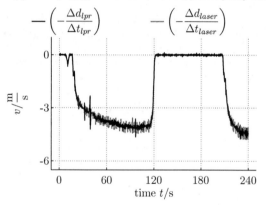

(b) Velocity calculated from the change in distance.

Figure 5.17: Airfield, velocity measurements: The bicycle is accelerated to velocities of up to 4.5 m/s. (a) The velocity measured by the LPR v_{meas} matches the ratio of the change in distance between two consecutive laser measurements Δd_{laser} and the time between those measurements Δt_{laser}. Since both stations are moving apart from each other the velocity is considered to be negative. (b) The velocity of the bicycle can also be calculated from the ratio of the change in distance between two consecutive LPR measurements Δd_{lpr} and the time between those measurements Δt_{lpr}. Again, the measurement results of the LPR match the results obtained by the laser ranging system.

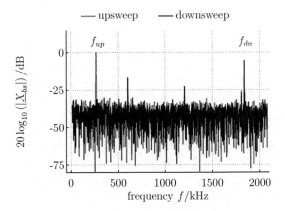

Figure 5.18: Airfield, maximum range of the system (measured data, base station): The LPR can be used to measure the distance between the base station and the transponder over the entire length of the runway of 2.2 km. The maximum range of the system can be estimated from the spectrum of the low-pass filtered mixed signal at a distance of approximately 2 km. The level measured during the downsweep (at f_{dn}, black line) is smaller than the level measured during the upsweep (at f_{up}, gray line), since f_{dn} is close to the cut-off frequency of the low-pass filter. However, the signals can be attenuated by at least another 6 dB, implying a range of more than 4 km.

caused by the power supply at all times. However, measurement errors can be detected by monitoring the velocity measurements. If the wrong peaks are detected the velocity measurements usually exceed plausible values and velocities larger than 100 m/s are measured.

In figure 5.18 the frequency of the low-pass filtered mixed signal can still be detected reliably during both measurement sweeps if the level is reduced by another 6 dB. Since the spectra were recorded at a distance of 2 km this implies a total transmission range of the system of approximately 4 km. This is a good match with the results shown in table 5.3. In practice, the measurement range is limited to 2.23 km due to the frequency the low-pass filtered mixed signal is digitized with in the base station (4.27).

5.3 Measurements in the Hallway of an Office Building

While the airfield tests demonstrated the performance of the measurement system at hand in unobstructed free-space conditions, behavior in an indoor environment is evaluated by measurements in an office building. Figure 5.19 depicts the measurement setup in the hallway of an office building.

The base station is connected to an antenna with a directional gain of 12 dBi [37]. It is placed on a table along with the laser ranging system that has already been used

base station antenna
transponder antenna

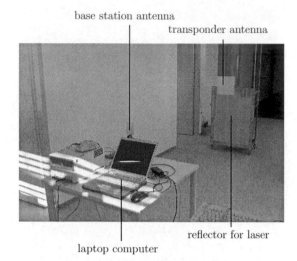

reflector for laser
laptop computer

(a) Shortest distance between the radar stations.

transponder station

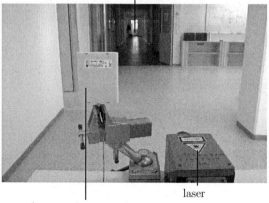

laser
base station antenna

(b) Longest distance between the radar stations.

Figure 5.19: Hallway, setup: The base station and a laser ranging system are mounted on a table (a, left hand side). The transponder and the reflector for the laser ranging system are mounted on a cart (a, right hand side). The cart can be moved along the hallway up to a distance of 26 m (b). A laptop computer is used to record the distance measurements of the LPR and the laser ranging system.

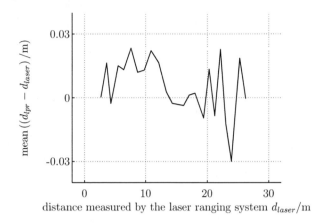

distance measured by the laser ranging system d_{laser}/m

Figure 5.20: Hallway, mean of the distance measurements: The distance between both stations
is increased from 3 m to 26 m in steps of 1 m. 200 measurements are taken at each
position. The mean difference between the distances measured by the LPR d_{lpr}
and the laser ranging system d_{laser} is below 30 mm at all points. Note, that the
constant offset due to the length of the antenna cables has been subtracted from
the distances measured by the LPR as well.

at the airfield. The transponder is connected to an antenna with a directional gain of
23 dBi [38]. The transponder and a reflector for the laser ranging system are mounted
on a cart. The cart is moved along the hallway and the distance between the base
station and the transponder can be increased to up to 26 m.

For the measurement results shown in figure 5.20 and figure 5.21 the distance between
the radar stations is increased from 3 m to 26 m in steps of approximately 1 m. At each
position 200 measurements are acquired by the LPR before the cart is moved again.
Furthermore, the laser ranging system is used to measure the distance to the reflector.
As before the system is calibrated prior to the measurements and the offset due to the
electrical length of the antenna cables is compensated for at a known distance.

Figure 5.20 depicts the mean difference of the distances measured by the LPR and
the laser ranging system. It is below 30 mm at all distances. The results are similar to
the results obtained at the airfield.

The standard deviation of the distance measurements is shown in figure 5.21. It is
less than 9 mm at all times, which is slightly higher than the standard deviation of 7 mm
measured at the airfield. However, the difference is considered to be insignificant since
the deviation of the clock frequencies has not been monitored during the measurements.
If the mismatch of the sweep rates of both radar stations changes slightly the standard
deviation can increase easily by a few millimeters as shown in figure 5.9.

Finally, the overall error of the distance measurements of the LPR with respect to
the laser ranging system is calculated. It is defined as the difference of the distances estimated by the LPR d_{lpr} and the laser ranging system d_{laser}. All distance measurements

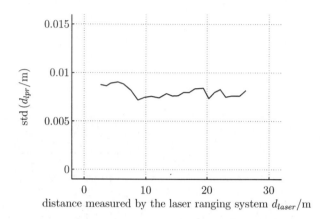

Figure 5.21: Hallway, standard deviation of the distance measurements: The standard deviation of the distances measured by the LPR d_{lpr} is below 9 mm for all measurements at distances between 3 m and 26 m.

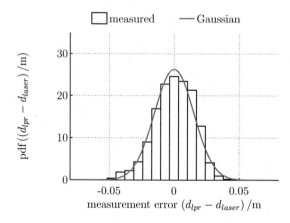

Figure 5.22: Hallway, overall error of the distance measurements: The overall error is defined as the difference between the distances measured by the LPR d_{lpr} and the distances measured by the laser ranging system d_{laser}. The measurement results obtained over the entire range are taken into account. The probability density function (pdf) of the overall error approximates a Gaussian distribution. The standard deviation is only 15 mm.

acquired at each position are taken into account. The probability density function of the overall error is close to a Gaussian distribution. It is shown in figure 5.22. The standard deviation of the overall error of 15 mm matches the standard deviation obtained during the measurements at the airfield.

Again, note that both measurement errors of the laser ranging system and the LPR contribute to the overall error. If the reflector and the antennas at the transponder side are not perpendicular to the line-of-sight to the base station the laser ranging system can measure a different distance than the LPR. Therefore, a more precise test setup is required to evaluate the system performance. It is presented in the next section.

5.4 Test Setup with a Linear Unit in an Indoor Environment

For the final indoor test both radar stations are connected to antennas with a horizontal and vertical 3 dB beam width of 9 ° [38]. The low beam width of the antennas minimizes multipath distortions from the environment. The base station antenna is mounted on a linear unit. The antenna of the transponder is placed approximately 2 m in front of the linear unit. The measurement setup is shown in figure 5.23. In this setup the measured distance between the stations is the sum of the constant electrical length of the antenna cables, a constant offset of approximately 2 m due to the spacing between the antenna of the transponder and the beginning of the linear unit, and the position of the tray of the linear unit, i. e. the distance the antenna of the base station has been moved along the linear unit.

The distance between the antennas of the base station and the transponder is increased in steps of 2 mm. At each position 1000 samples of the distance are acquired by the LPR before the linear unit is moved to the next position. The constant offsets are subtracted from the obtained distance measurements and only the measurement error, i. e. the difference between the true position d_{lu} of the tray of the linear unit and the position d_{lpr} measured by the LPR, remains.

Since the true position of the tray of the linear unit is known with an inaccuracy of less than 1 mm, the measurement error is assumed to be caused by the LPR only. The exact knowledge of the position of the tray of the linear unit is the main advantage over the measurement setup presented in section 5.3. There, measurement errors of the LPR as well as errors of the laser ranging system contributed to the overall measurement error.

Figure 5.24(a) depicts the mean measurement error of the LPR with respect to the true position of the linear unit. It is within ±15 mm over the entire range for all positions of the tray of the linear unit between 0 m and 4 m. The error is caused mainly by short multipath signal components that cannot be separated from the line-of-sight path.

In theory, the resolution of multipath components would be improved if the bandwidth of the radar signals was increased. This has been shown in section 4.9.2. In practice the bandwidth of the radar sweeps is limited to 132 MHz, since a bandwidth of

transponder antenna

base station antenna

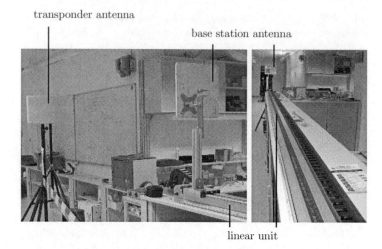

linear unit

Figure 5.23: Linear unit, setup: The antenna of the base station is mounted on a linear unit. The antenna of the transponder is located approximately 2 m in front of the linear unit. The position of the antenna of the base station on the linear unit is increased from 0 m to 4 m in steps of 2 mm. Hence, the distance between the antennas increases from 2 m to 6 m. 1000 measurements are taken at each point.

only 150 MHz is available in the ISM band at 5.8 GHz and some bandwidth is reserved for FSK communication.

The standard deviation of the distance measurements of the LPR is shown in figure 5.24(b). It is approximately 7 mm at all positions of the linear unit. The standard deviation of 7 mm matches the standard deviation obtained during the delay line measurements in section 5.1.1.

Finally, the overall error of the distance measurements is calculated. It is defined as the difference of the distances d_{lpr} measured by the LPR and the true position of the linear unit d_{lu}. All measurements obtained over the entire length of the linear unit are taken into account. Once again, the probability density function of the overall measurement error approximates a Gaussian distribution. It is shown in figure 5.25.

The standard deviation of the overall error is only 8.4 mm. This is significantly smaller than the standard deviation obtained during the airfield measurements or the hallway measurements since the measurement errors of the laser ranging system do not contribute to the overall error anymore.

5.5 Summary

In this chapter the results of an extensive measurement campaign have been presented. A delay line has been used to evaluate the performance of the radar system without

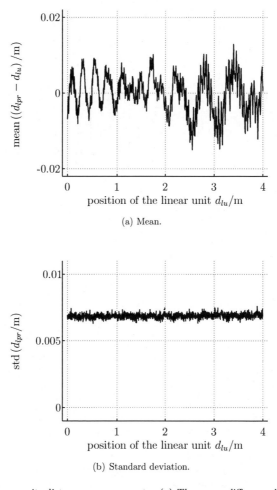

(a) Mean.

(b) Standard deviation.

Figure 5.24: Linear unit, distance measurements: (a) The mean difference between the distances measured by the LPR d_{lpr} and the position of the linear unit d_{lu} is within 15 mm at all positions. Note, that the constant offset due to the length of the antenna cables and the position of the transponder antenna has been subtracted from the distances measured by the LPR as well. (b) The standard deviation of the distances measured by the LPR d_{lpr} is approximately 7 mm over the entire range.

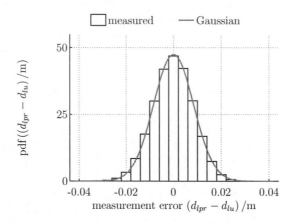

Figure 5.25: Linear unit, overall error of the distance measurements: The overall error is defined as the difference between the distances measured by the LPR d_{lpr} and the position of the linear unit d_{lu}. The measurement results obtained over the entire range are taken into account. The probability density function (pdf) of the overall error approximates a Gaussian distribution. The standard deviation is only 8.4 mm.

distortions due to the environment. If multipath propagation is neglected the distance and velocity measurements are subjected to a Gaussian distribution.

The standard deviation of the distance and velocity measurements is as low as 6.4 mm and 0.16 m/s, respectively. The minimum standard deviations are obtained only if the sweep rates of the base station and the transponder match and the SNR of the low-pass filtered mixed signal is determined by the phase noise of the PLL in the signal generator of each station. The standard deviation of the distance and velocity measurements increases to up to 30 mm and 0.7 m/s, respectively, if the deviation of the clock frequencies of both radar stations is increased from 0 ppm to 50 ppm. These measurement results are conclusive with the simulation results obtained in section 4.7.

The measurement range where the SNR of the low-pass filtered mixed signal is determined by the phase noise of the PLL depends on the directional gain of the antennas used with the system. If antennas with a directional gain of 23 dBi are used the minimum standard deviation of the distance measurements is achieved for distances below 200 m. For larger distances the SNR of the low-pass filtered mixed signal is not determined by the phase noise of the PLL anymore. Then, the standard deviation of the distance measurements increases to up to 30 mm at the maximum measurement range of 2.23 km. The maximum measurement range of the current hardware setup is limited by the frequency the low-pass filtered mixed signal is sampled with in the base station (4.27).

The radar distance measurement system has been tested at an airfield over distances of more than 550 m. The measurement results are compared to the distance measurements of a laser ranging system. The mean deviation of the distances measured by both systems is within ±30 mm over the entire range. The probability density function of the overall error approximates a Gaussian distribution with a standard deviation of only 15 mm.

Furthermore, the measurement system has been tested in an indoor environment. Here, a linear unit has been used to precisely control the distance between the antennas of the base station and the transponder to avoid any errors caused by the laser ranging system. In this setup, the probability density function of the overall error of the radar system presented in this thesis with respect to the position of the linear unit approximates a Gaussian distribution. The standard deviation is only 8.4 mm. The mean deviation of the true position of the linear unit and the position measured by the LPR is within ±15 mm for all positions of the linear unit.

The measurement results prove the outstanding performance of the measurement system presented in this thesis and the underlying algorithms for synchronization and distance and velocity measurement which have been derived in chapter 2.

Chapter 6 — Conclusion and Outlook

The thesis at hand provides a detailed analysis of a novel secondary radar concept for precise distance and velocity measurement. The measurement system consists of at least two radar stations, a base station and a transponder. FMCW radar sweeps are used to synchronize the transponder to the base station with high precision. After synchronization, a synchronized reply, i. e. another FMCW radar signal, is sent back to the base station and the distance and relative velocity of the stations are measured.

During synchronization, a local signal in the transponder is adjusted to a signal received from the base station. The local signal is multiplied with the received signal. Subsequently, the offsets in time and in frequency of both signals are calculated from the frequency of the low-pass filtered mixed signal in the transponder. Similarly, the distance and relative velocity of the radar stations are calculated from the frequency of the low-pass filtered mixed signal in the base station during the measurement sweeps.

A thorough mathematical analysis of the underlying algorithms for synchronization and distance measurement has been given. For the first time the effect of the relative motion of the radar stations during the entire synchronization and measurement process has been taken into account as well. It has also been shown how the Doppler frequency shift of the radar signals can be used to measure the relative velocity of the stations.

Furthermore, a novel iFDMA scheme for the synchronized signals transmitted by multiple transponders has been derived. The method allows to measure the distance and relative velocity of the base station and multiple transponders with a single measurement upsweep and downsweep. Thereby, the measurement rate in 2D scenarios has been improved greatly compared to previous TDMA approaches.

It has also been shown that the velocity measurements provide a means to detect aliases of the frequency of the signals evaluated in the base station. The aliases occur if the distance between the stations increases beyond a maximum range which is given by the frequency the signals are digitized with in the base station. It is shown how the aliases can be detected and corrected if the parameters of the radar signals are chosen properly. Thereby, a longer measurement range of the system is enabled.

Based on the algorithms for synchronization and measurement a measurement system has been designed. The system consists of two active radar units which share a common hardware setup. However, the first unit is configured to work as a base station. It measures its distance to the second station which is set to transponder mode.

The main focus of the thesis at hand is on the evaluation of the performance of the measurement system and the identification of sources of error. Simulations and measurements have been used to investigate the dependence of the measurement results on various influences. The most important sources of error are multipath distortions, the mismatch of the sweep rates of the base station and the transponder, the SNR of the low-pass filtered mixed signal, and finally the deviation of the sampling frequency and the phase velocity of the radar signals from their respective nominal values.

The effect of multipath propagation has been studied in a comprehensive manner. It has been concluded that multipath distortions result in multiple peaks in the power spectral densities used to estimate the frequency of the low-pass filtered mixed signal during the synchronization and measurement sweeps. It has been shown that it is always possible to identify the peak corresponding to the line-of-sight path if the peaks corresponding to the line-of-sight and non-line-of-sight components can be resolved. The resolution of the radar system improves as the bandwidth of the sweeps is increased.

If multipath distortions are neglected the mismatch of the sweep rates of the base station and the transponder is the dominant source of error. It has been shown by simulations and measurements that the peaks in the power spectral densities widen, if the sweep rates do not match, since the frequency of the low-pass filtered mixed signal changes linearly during the sampling time. Consequently, the estimation of the frequency becomes more inaccurate and the standard deviation of the distance and velocity measurements increases. However, the sweep rate of the transponder can be adjusted to the sweep rate of the base station by proper hardware design, i. e. if a voltage controlled crystal oscillator is used to clock the stations.

It has also been shown that the standard deviation of the measurement results depends on the SNR of the low-pass filtered mixed signal. The minimum standard deviation is achieved for short distances. Here, the SNR is determined by the phase noise of the hardware used to generate the FMCW radar signals. It has been pointed out that the SNR of the low-pass filtered mixed signal becomes dependent on the distance of the radar units if the distance increases beyond a certain threshold, which depends on the gain of the antennas used with the system. If the distance of the radar stations exceeds the threshold, the SNR of the low-pass filtered mixed signal is determined by the path-loss during the transmission of the FMCW radar signals. Consequently, the SNR is reduced as the distance of the radar stations increases.

Furthermore, it has been concluded that precise knowledge of constant system parameters, e. g. the frequency the low-pass filtered mixed signal is digitized with or the phase velocity of the radar signals, is mandatory for very high precision measurements. Any deviation from the respective nominal values results in a measurement error proportional to the measured distance and the deviation of the constant system parameters. Since the measurement error is proportional to the measured distance, it can be compensated for if the measurement system is calibrated at a precisely known distance.

The measurement system presented in this thesis works in the ISM band at 5.8 GHz where a bandwidth of 150 MHz is available. It has been tested under various conditions. The distance measurement error has been found to be a Gaussian random variable if a single line-of-sight connection is assumed between the stations. It has been shown that the minimum standard deviation of the distance measurements is 6.4 mm. It can be achieved for distances below 200 m if the sweep rates of the radar stations match and antennas with a gain of 23 dBi are used with the system.

However, for the current hardware setup the mismatch of the sweep rates is proportional to the deviation of the clock frequencies of the radar stations. The clock frequencies can differ by up to 50 ppm and the maximum mismatch of the sweep rates is 100 ppm. At the maximum mismatch of the sweep rates the standard deviation of the distance measurements is 30 mm.

The system at hand has been tested successfully in practice for a range of up to 2.2 km. An analysis of the signal level even implies a possible transmission range of more than 4 km if antennas with a gain of 23 dBi are used. However, the measurement range is limited to 2.23 km due to the frequency the low-pass filtered mixed signal is digitized with in the base station and the computational power and the memory restrictions of the DSP.

The results of the thesis at hand imply many interesting topics for future research. For instance, it is interesting to investigate how the widening of the peaks due to the mismatch of the sweep rates of the base station and the transponder affects the resolution of short multipath components.

Furthermore, a linear modulation of the FMCW radar signals has been assumed for the analysis of the synchronization and measurement principle in this thesis. Non-linearities in the FMCW signals will undoubtedly impair the performance of the measurement system. Therefore, their effect on the accuracy of the synchronization and measurement results must be studied.

Furthermore, it has been shown that the resolution of multipath components can be improved if the bandwidth of the sweeps is increased. Therefore, Ultra-WideBand (UWB) systems promise an optimized performance especially in dense multipath environments. Waldmann et al. developed a UWB system working at frequencies from 7 GHz to 8 GHz [104–106]. For the UWB system the algorithms presented in this thesis are modified slightly. For instance, the sweeps used for synchronization and distance measurement are superimposed with short pulses to fulfill the emission limits given by regulatory authorities, e.g. the Federal Communications Commission (FCC) [21]. Initial measurement results promise a good resolution of multipath components due to the high sweep bandwidth of 1 GHz.

Yet another implementation of the secondary radar concept presented in this thesis is working in the ISM band at 61 GHz where a bandwidth of 500 MHz is available. Measurement results obtained in an indoor environment also indicate a good resolution of multipath components due to the high bandwidth of the FMCW radar signals [71,72].

It is also interesting to investigate possibilities to reduce the power consumption of the system. The calculation of the FFT of the low-pass filtered mixed signal can be implemented in an FPGA. Thereby, the time required for the calculations can be reduced. If the measurement rate is not increased the system can be powered down for the remaining time of the measurement cycle. Power can also be saved by redesigning the signal generator, e.g. by replacing the direct digital synthesizer by a fractional-N synthesizer.

Finally, chip integration of the entire system promises a reduction of the power consumption, the size, and the price of the measurement system. The system presented in this thesis is currently used for industrial applications like crane collision avoidance or tracking of vehicles. However, smaller and less expensive units enable new applications, e.g. mobile personal information and navigation systems, location based services, or security applications.

Bibliography

[1] Abracon Corporation, "ASTX-H08," October 2008. [Online]. Available: http://www.abracon.com

[2] E. S. Alvarez, "General discussion of geometrical factors affecting trilateration solutions for position and velocity," U.S. Army Test and Evaluation Command, Range Instrumentation Systems Office, Tech. Rep., 1966.

[3] Analog Devices, Inc., "A technical tutorial on digital signal synthesis," Tech. Rep., 1999.

[4] Analog Devices, Inc., "AD9852 CMOS 300 MSPS complete DDS," May 2007. [Online]. Available: http://www.analog.com

[5] Analog Devices, Inc., "AD9954 direct digital synthesizer," 2007. [Online]. Available: http://www.analog.com

[6] Anritsu GmbH, *Series MG369XB Synthesized Signal Generators Operation manual*, March 2007. [Online]. Available: http://www.anritsu.de

[7] D. Banerjee, *PLL Performance, Simulation and Design*, 4th ed. Dog Ear Publishing, LLC, August 2006.

[8] D. K. Barton and S. A. Leonov, *Radar Technology Encyclopedia*. Norwood: Artech House, February 1997.

[9] N. C. Beaulieu, "Introduction to "certain topics in telegraph transmission theory"," *Proceedings of the IEEE*, vol. 90, no. 2, pp. 276–279, February 2002.

[10] Bores Signal Processing, "The zoom FFT," Bores Signal Processing, Tech. Rep., 2007.

[11] E. O. Brigham, *The fast Fourier transform and its applications*. Englewood Cliffs: Prentice Hall, March 1988.

[12] I. N. Bronstein, K. A. Semendjajew, G. Musiol, and H. Mühlig, *Taschenbuch der Mathematik*, 5th ed. Frankfurt am Main: Verlag Harri Deutsch, 2001.

[13] H. Chernoff and E. L. Lehmann, "The use of maximum likelihood estimates in χ^2 tests for goodness of fit," *Annals of Mathematical Statistic*, vol. 25, no. 3, pp. 579 – 586, 1954.

[14] J. W. Cooley and J. W. Tukey, "An algorithm for the machine calculation of complex Fourier series," *Mathematics of Computation*, vol. 19, pp. 297–301, 1965.

[15] C. M. Crain, "The dielectric constant of several gases at a wave-length of 3.2 centimeters," *Physical Review*, vol. 74, no. 6, pp. 691–693, September 1948.

[16] R. J. Crinon, "Sinusoid parameter estimation using the fast Fourier transform," in *IEEE International Symposium on Circuits and Systems*, vol. 2, Portland, USA, May 1989, pp. 1033–1036.

[17] P. Daly, "Navstar GPS and GLONASS: global satellite navigation systems," *Electronics and Communication Engineering Journal*, vol. 5, no. 6, pp. 349–357, December 1993.

[18] H. Dodel and D. Häupler, *Satellitennavigation. GALILEO, GPS, GLONASS, integrierte Verfahren.* Hüthig Telekommunikation, 2004.

[19] H. Elmer and G. Magerl, "Hochauflösende Ultraschall-Entfernungsmessung für große Distanzen," in *XVI. Messtechnisches Symposium des AHMT e.V.*, October 2002, pp. 137–148. [Online]. Available: http://www2.mst.ei.tum.de/ahmt/publ/symp/2002/2002_137.pdf

[20] elspec GmbH, "LL2773-AF: Koaxiales HF-Kabel," June 2008. [Online]. Available: http://www.elspec.de/

[21] Federal Communications Commission, "First report and order (FCC 02-48, ET Docket 98-153)," February 2002.

[22] K. Feher, *Wireless Digital Communications.* Englewood Cliffs, NJ: Prentice Hall, 1995.

[23] J. C. Fuentes Michel, "Non-linear observation, filtering and smoothing in markovian processes applied to wireless positioning of moving objects with high dynamics," Ph.D. dissertation, Clausthal University of Technology, 2009.

[24] J. C. Fuentes Michel, H. Millner, and M. Vossiek, "A novel wireless forklift positioning system for indoor and outdoor use," in *5th Workshop on Positioning, Navigation and Communication*, Hannover, March 2008, pp. 219–227.

[25] P. Gerdsen and P. Kröger, *Digitale Signalverarbeitung in der Nachrichtenübertragung - Elemente, Bausteine, Systeme und ihre Algorithmen*, 2nd ed. Berlin: Springer Verlag, September 1996.

[26] R. Gierlich, J. Hüttner, A. Dabek, and M. Huemer, "Performance analysis of FMCW synchronization techniques for indoor radiolocation," in *2007 European Conference on Wireless Technologies*, Munich, Germany, October 2007, pp. 24–27.

[27] E. P. Glennon, A. G. Dempster, and C. Rizos, "Feasibility of air target detection using GPS as a bistatic radar," *Positioning (Journal of Global Positioning Systems)*, vol. 5, no. 1, pp. 119–126, December 2006.

[28] S. A. Golden and S. S. Bateman, "Sensor measurements for Wi-Fi location with emphasis on time-of-arrival ranging," *IEEE Transactions on Mobile Computing*, vol. 6, no. 10, pp. 1185–1198, October 2007.

[29] J. B. Groe and L. E. Larson, *CDMA Mobile Radio Design*. Boston, London: Artech House, 2000.

[30] A. G. Götz, "Detektionsalgorithmik und Mehrwegekompensationsverfahren für Local-Positioning-Radar-Systeme," Master's thesis, Friedrich-Alexander-Universität Erlangen-Nürnberg, May 2008.

[31] P. Gulden, "Anwendung und Optimierung von neuartigen, State-Space-basierten Frequenzschätzverfahren für hochauflösende FMCW-Radarsensoren," Master's thesis, Universität GHS Siegen, 1998.

[32] P. Gulden, *System Concepts for Novel Optical Distance Sensors Based on Inherently-Mixing Detectors*. Düsseldorf: VDI Verlag, 2003.

[33] P. Gulden, "Wireless vehicle localization for industrial applications," Workshop, June 2008, 2008 IEEE MTT-S International Microwave Symposium.

[34] F. Gustafsson, F. Gunnarsson, N. Bergman, U. Forssell, J. Jansson, R. Karlsson, and P.-J. Nordlund, "Particle filters for positioning, navigation, and tracking," *IEEE Transactions on Signal Processing*, vol. 50, no. 2, pp. 425–437, February 2002.

[35] F. J. Harris, "On the use of windows for harmonic analysis with the discrete Fourier transform," *Proceedings of the IEEE*, vol. 66, no. 1, pp. 51–83, January 1978.

[36] J. Hightower and G. Borriello, "Location systems for ubiquitous computing," *Computer*, vol. 34, no. 8, pp. 57–66, August 2001.

[37] Huber+Suhner, "Planar antenna for wireless communication SPA 5600/65/12/0/V," August 2006. [Online]. Available: http://www.hubersuhner. com

[38] Huber+Suhner, "Planar antenna for wireless communication SPA 5600/9/23/0/V," February 2007. [Online]. Available: http://www.hubersuhner. com

[39] International Telecommunication Union, "Radio regulations edition of 2008," September 2008. [Online]. Available: http://www.itu.int/publ/R-REG-RR/en

[40] K. Kaemarungsi and P. Krishnamurthy, "Modeling of indoor positioning systems based on location fingerprinting," in *Twenty-third Annual Joint Conference of the IEEE Computer and Communications Societies INFOCOM 2004*, vol. 2, March 2004, pp. 1012–1022.

[41] Y. Kameda, T. Takemasa, and Y. Ohta, "Outdoor see-through vision utilizing surveillance cameras," in *Third IEEE and ACM International Symposium on Mixed and Augmented Reality*, November 2004, pp. 151–160.

[42] S. M. Kay, *Modern Spectral Estimation: Theory and Application.* Englewood Cliffs: Prentice Hall, 1987.

[43] S. M. Kay and A. K. Shaw, "Frequency estimation by principal component AR spectral estimation method without Eigendecomposition," *IEEE Transactions on Acoustics, Speech and Signal Processing*, vol. 36, no. 1, pp. 95–101, January 1988.

[44] H. Klausing and W. Holpp, *Radar mit realer und synthetischer Apertur: Konzeption und Realisierung.* Oldenbourg, October 1999.

[45] M. Kossel, H. R. Benedickter, R. Peter, and W. Bächtold, "Microwave backscatter modulation systems," in *2000 IEEE MTT-S International Microwave Symposium*, Boston, USA, June 2000, pp. 1427–1430.

[46] KVG Quartz Technology GmbH, "Email regarding the phase noise and clock jitter of the XO-9000 series," November 2008.

[47] KVG Quartz Technology GmbH, "XO-9000 series," Oktober 2008. [Online]. Available: http://www.kvg-gmbh.de

[48] Larsen Antennas, "5.73-5.87 GHz Radome Omni - 6 dBi omnidirectional antenna," January 2009. [Online]. Available: http://www.larsen-antennas.com

[49] LASE GmbH, *Laser-Distanzmesser ELD P-Serie*, January 2004. [Online]. Available: http://www.lase.de

[50] J. Latvala, J. Syrjärinne, H. Ikonen, and J. Niittylahti, "Evaluation of RSSI-based human tracking," in *2000 European Signal Processing Conference*, vol. 4, 2000, pp. 2273–2276.

[51] H. Liu, H. Darabi, P. Banerjee, and J. Liu, "Survey of wireless indoor positioning techniques and systems," *IEEE Transactions on Systems, Man, and Cybernetics, Part C: Applications and Reviews*, vol. 37, no. 6, pp. 1067–1080, November 2007.

[52] D. G. C. Luck, *Frequency modulated radar.* McGraw-Hill, 1949.

[53] M. R. Mahfouz, C. Zhang, B. C. Merkl, M. J. Kuhn, and A. E. Fathy, "Investigation of high-accuracy indoor 3-d positioning using UWB technology," *IEEE Transactions on Microwave Theory and Techniques*, vol. 56, no. 6, pp. 1316–1330, June 2008.

[54] S. Max, "Neuartige hybride Ortungsverfahren basierend auf synthetischen Aperturen zur Schätzung der Position und Ausrichtung von Transpondern im Raum," Ph.D. dissertation, Technische Universität Clausthal, 2008.

[55] Maxim Integrated Products, "Clock jitter and phase noise conversion," September 2004. [Online]. Available: http://pdfserv.maxim-ic.com/en/an/AN3359.pdf

[56] A. K. L. Miu, "Design and implementation of an indoor mobile navigation system," Master's thesis, Massachusetts Institute of Technology, January 2002.

[57] MtronPTI, "MtronPTI's oscillator jitter basics," Tech. Rep., November 2008. [Online]. Available: http://www.mtronpti.com/pdf/contentmgmt/oscillator_jitter_basics-2.pdf

[58] Narda Microwave East, "Thumb wheel and panel mount step attenuators," June 2008. [Online]. Available: http://www.nardamicrowave.com

[59] W. Navidi, W. S. Murphy, and W. Hereman, "Statistical methods in surveying by trilateration," *Computational Statistics and Data Analysis*, vol. 27, no. 2, pp. 209–227, April 1998.

[60] B. Neubig and W. Briese, *Das große Quarzkochbuch*. Feldkirchen: Franzis-Verlag, 1997.

[61] A. Nuttall, "Some windows with very good sidelobe behavior," *IEEE Transactions on Acoustics, Speech and Signal Processing*, vol. 29, no. 1, pp. 84–91, February 1981.

[62] H. Nyquist, "Certain topics in telegraph transmission theory," *Transactions of the AIEE*, vol. 47, pp. 617–644, February 1928, reprinted in "Proceedings of the IEEE", vol, 90, no. 2, pp. 280-305, February 2002.

[63] N. Patwari, J. N. Ash, S. Kyperountas, A. O. Hero III, R. L. Moses, and N. S. Correal, "Locating the nodes: Cooperative localization in wireless sensor networks," *IEEE Signal Processing Magazine*, vol. 22, no. 4, pp. 54–69, July 2005.

[64] W. E. Phillips, "The permittivity of air at a wavelength of 10 centimeters," *Proceedings of the IRE*, vol. 38, no. 7, pp. 786–790, July 1950.

[65] J. G. Proakis, *Digital Communications*, 4th ed. Singapore: McGraw-Hill, 2001.

[66] B. D. Rao and K. S. Arun, "Model based processing of signals: a state space approach," *Proceedings of the IEEE*, vol. 80, no. 2, pp. 283–309, February 1992.

[67] T. S. Rappaport, *Wireless Communications: Principles and Practice*, ser. Communications Engineering and Emerging Technologies. Prentice Hall, January 2002.

[68] Regulierungsbehörde für Telekommunikation und Post, "Frequenznutzungsplan gemäß TKG über die Aufteilung des Frequenzbereichs von 9kHz bis 275GHz," April 2008. [Online]. Available: http://www.bundesnetzagentur.de/media/ archive/13358.pdf

[69] S. Röhr, "Round-trip time-of-flight Lokalisierungssystem mit neuartigem Systemkonzept für universelle Lokalisierungsmodule," Master's thesis, Technische Universität Dresden, March 2005.

[70] S. Röhr, *Neuartiges Sekundärradar zur Entfernungsmessung - Konzept, Analyse, Erprobung.* Saarbrücken: VDM Verlag Dr. Müller, April 2008.

[71] S. Röhr, "Präzise Kranortung mit Millimeterwellen," *Hebezeuge - Fördermittel*, vol. 2009, no. 9, pp. 436 – 438, September 2009.

[72] S. Röhr and P. Gulden, "Hochpräzise Kranortung mit Millimeterwellen," in *17. Internationale Kranfachtagung*, Dresden, Germany, March 2009, pp. 85–104.

[73] S. Röhr, P. Gulden, and M. Vossiek, "Method for high precision clock synchronization in wireless systems with application to radio navigation," in *2007 IEEE Radio and Wireless Symposium*, January 2007, pp. 551–554.

[74] S. Röhr, P. Gulden, and M. Vossiek, "Novel secondary radar for precise distance and velocity measurement in multipath environments," in *2007 European Radar Conference*, October 2007, pp. 182–185.

[75] S. Röhr, P. Gulden, and M. Vossiek, "Precise distance and velocity measurement for real time locating in multipath environments using a frequency modulated continuous wave secondary radar approach," *IEEE Transactions on Microwave Theory and Techniques*, vol. 56, no. 10, pp. 2329–2339, October 2008.

[76] S. Röhr, M. Vossiek, and P. Gulden, "Method for high precision radar distance measurement and synchronization of wireless units," in *2007 IEEE MTT-S International Microwave Symposium*, June 2007, pp. 1315–1318.

[77] D. C. Rife, "Digital tone parameter estimation in the presence of Gaussian noise," Ph.D. dissertation, Polytechnic Institute of Brooklyn, 1973.

[78] D. C. Rife and R. R. Boorstyn, "Single-tone parameter estimation from discrete-time observations," *IEEE Transactions on Information Theory*, vol. 20, no. 5, pp. 591–598, September 1974.

[79] Rohde & Schwarz GmbH & Co. KG, *Operating Manual: Spectrum Analyzer FSP*, November 2006. [Online]. Available: http://www.rohde-schwarz.com

[80] H. Rudolph, "OCXOs - die Königsklasse," *Elektronik - Fachzeitschrift für industrielle Anwender und Entwickler*, vol. 57, no. 23, pp. 64–68, November 2008.

[81] R. O. Schmidt, "Multiple emitter location and signal parameter estimation," *IEEE Transactions on Antennas and Propagation*, vol. 34, no. 3, pp. 276–280, March 1986.

[82] C. Seisenberger and M. Vossiek, "Verfahren und Vorrichtungen zur Synchronisation von Funkstationen und zeitsynchrones Funkbussystem," December 2003, Siemens AG, Patentschrift DE000010157931C2.

[83] J. S. Seybold, *Introduction to RF Propagation*. Wiley, September 2005.

[84] C. E. Shannon, "A mathematical theory of communication," *The Bell System Technical Journal*, vol. 27, pp. 379–423, 623–656, July, October 1948.

[85] M. I. Skolnik, *Radar Handbook*, 3rd ed. McGraw-Hill, January 2008.

[86] J. O. Smith, *Spectral Audio Signal Processing*, October 2008. [Online]. Available: http://ccrma.stanford.edu/~jos/sasp/

[87] M. R. Spiegel, *Schaum's Mathematical Handbook of Formulas and Tables*, 1st ed. McGraw-Hill, February 1968.

[88] S. Steiniger, M. Neun, and A. Edwardes, "Foundations of location based services," Retrieved online in May 2009, 2006. [Online]. Available: http://www.geo.unizh.ch/publications/cartouche/lbs_lecturenotes_steinigeretal2006.pdf

[89] A. Stelzer, M. Pichler, and S. Schuster, "Frequency estimation in linear frequency modulated radar systems," *Journal of RF-Engineering and Telecommunications (FREQUENZ)*, vol. 60, no. 1-2, pp. 31–36, January 2006.

[90] A. Stelzer, K. Pourvoyeur, and A. Fischer, "Concept and application of LPM - a novel 3-D local position measurement system," *IEEE Transactions on Microwave Theory and Techniques*, vol. 52, no. 12, pp. 2664–2669, December 2004.

[91] A. Stelzer, K. Pourvoyeur, and P. Scherz, "Multi sensor data fusion in local positioning," workshop, June 2008, 2008 IEEE MTT-S International Microwave Symposium.

[92] G. W. Stimson, *Introduction to Airborne Radar*, 2nd ed. SciTech Publishing, January 1998.

[93] A. G. Stove, "Linear FMCW radar techniques," *IEE Proceedings F Radar and Signal Processing*, vol. 139, no. 5, pp. 343–350, October 1992.

[94] Symeo GmbH, "LPR Antennen: Montage und Anschluss," March 2008. [Online]. Available: http://www.symeo.com/English/index.html

[95] Texas Instruments Inc., "C28x FFT library," May 2002. [Online]. Available: http://www-s.ti.com/sc/psheets/sprc081/sprc081.zip

[96] J. Thornton and D. J. Edwards, "Range measurement using modulated retro-reflectors in FM radar system," *IEEE Microwave and Guided Wave Letters*, vol. 10, no. 9, pp. 380–382, September 2000.

[97] D. Tse and P. Viswanath, *Fundamentals of Wireless Communication*. Cambridge: Cambridge University Press, May 2005.

[98] A. Urruela, J. Sala, and J. Riba, "Average performance analysis of circular and hyperbolic geolocation," *IEEE Transactions on Vehicular Technology*, vol. 55, no. 1, pp. 52–66, January 2006.

[99] M. Vossiek and P. Heide, "Studie über Verfahren und Algorithmen zur präzisen Abstands- und Füllstandsmessung mit Radar," June 1997.

[100] M. Vossiek, P. Heide, M. Nalezinski, and V. Magori, "Novel FMCW radar system concept with adaptive compensation of phase errors," in *26th European Microwave Conference*, vol. 1, Prague, Czech Republic, October 1996, pp. 135–139.

[101] M. Vossiek, L. Wiebking, P. Gulden, J. Wieghardt, and C. Hoffmann, "Wireless local positioning - concepts, solutions, applications," in *2003 Radio and Wirless Conference*, August 2003, pp. 219–224.

[102] M. Vossiek and S. Röhr, "Basics of wireless local positioning," in *Workshop at 2007 IEEE MTT-S International Microwave Symposium*, June 2007.

[103] M. Vossiek, L. Wiebking, P. Gulden, J. Wieghardt, C. Hoffmann, and P. Heide, "Wireless local positioning," *IEEE Microwave Magazine*, vol. 4, no. 4, pp. 77–86, December 2003.

[104] B. Waldmann, P. Gulden, M. Vossiek, and R. Weigel, "A pulsed frequency modulated ultra wideband technique for indoor positioning systems," *Journal of RF-Engineering and Telecommunications (FREQUENZ)*, vol. 62, no. 7-8, August 2008.

[105] B. Waldmann, P. Gulden, and M. Vossiek, "Pulsed frequency modulation techniques for high-precision ultra wideband ranging and positioning," in *2008 IEEE International Conference on Ultra-Wideband*, vol. 2, Hannover, Germany, September 2008, pp. 133–136.

[106] B. Waldmann, R. Weigel, and P. Gulden, "Method for high precision local positioning radar using an ultra wideband technique," in *2008 IEEE MTT-S International Microwave Symposium*, Atlanta, USA, 2008, pp. 117–120.

[107] L. Wiebking, *Entwicklung eines zentimetergenauen mehrdimensionalen Nahbereichs-Navigationssystems*. Düsseldorf: VDI Verlag, 2003.

[108] G. Xu, *GPS - Theory, Algorithms and Applications*. Springer, September 2007.